RECLAIM
RECYCLE
REUSE

and natural products
to help save the earth

Alan B Hayes

SALLY MILNER PUBLISHING

First published in 1992 by
Sally Milner Publishing Pty Ltd
67 Glassop Street
Birchgrove NSW 2041 Australia

© Alan Hayes, 1992

Production by Sylvana Scannapiego,
Island Graphics
Book design by Diana Kureen
Cover design by Shirley Peters
Illustrations by Shirley Peters
Typeset in Australia by Asset Typesetting Pty Ltd
Printed in Australia by Australian Print Group

National Library of Australia
Cataloguing-in-Publication data:

Hayes, Alan B. (Alan Bruce), 1949–   .
    Reclaim, recycle, reuse and natural ways to help
    save the earth.

    Includes index.
    ISBN 1 86351 065 6.

    1. Waste minimization. 2. Recycling (Waste, etc.).
    3. Natural products. I. Title. (Series: Milner healthy
    living guide).

363.737

CONTENTS

**To my family,
and all people who feel for
the health of our planet.**

INTRODUCTION

In past centuries the human race always turned to nature and the plants and earths found there to make medicines, cosmetics, soaps, and other beneficial products. They grew vegetables in harmony with nature so that a natural balance was achieved: plants protecting plants and the encouragement of friendly insects and animals to feed off the unfriendly insects.

Around the middle of the twentieth century all this changed. Chemicals began being poured onto the earth, sprayed into the air and used daily in homes, all to make life easier, crops more productive and the control of pests supposedly simpler and easier. It is now, as this century draws to a close, that human beings are realising the damage that has been caused, and will continue, by the mass use of dangerous chemicals, dioxins, phosphates and the like.

We have polluted the soil, our waterways and the air we breathe, all in the name of progress. A destructive pattern, that if not stopped, will completely destroy our environment and life on earth as we know it.

There is an alternative to the destructive chemicals: plant extracts that were used in our great grandparents' day are still a viable and safe substitute. These, combined with environmentally safe habits, must lead to a healthier planet that will guarantee future generations the same joy and beauty we have been able to experience.

Although it is important for all of us to work together globally and locally, it is just as necessary to take stock of the home front. You may ask what difference your personal contribution will make by recycling the kitchen scraps in a compost heap, or using homemade natural cosmetics. Yet if we all adopt a more responsible lifestyle, collectively it will eventually mean a difference.

This book takes you through many facets of environmentally sound practices that can be achieved on a domestic level. It tells you how to make safe, natural products for washing, cleaning, and polishing; ointments and other preparations for treating minor medical problems and first aid situations; cosmetics and toiletries for all the family; and simple laundry and pet products. It also includes many fragrant herb preparations to keep rooms and clothes smelling fresh and shows you how to extract both seed and herb oils.

Dozens and dozens of ideas and suggestions abound for recycling within the home, including how to make your own paper from recycled materials. Energy-efficient homes are discussed, along with inexpensive yet effective ways of using the sun and plants to naturally cool and warm your home. You are shown how to grow chemical-free vegetables and other plants, and establish a garden environment that is totally in harmony with nature.

This book has been written for people seeking an alternative to the chemicals that have invaded their lives. It's not just a recipe book, but a compendium to a more responsible lifestyle and healthier planet.

1 SOUND ENVIRONMENTAL PRACTICES

RECYCLING TO REDUCE WASTE

As a species, the human race produces about a billion tonnes of waste a year. And in many countries this waste is seen as a problem, rather than a resource that can be converted into something useful. Yet with public concern for our environment growing, there is hope for our future.

Every individual can have a positive effect on the environment. By recycling you help to:
- save energy
- save our natural resources
- assist employment
- stop littering
- save space at council garbage depots
- save our environment.

Most local councils in Australia now operate a recycling service for some items, such as newspapers and glass. Those items which they do not collect can usually be handled by recycling companies in your area. Check with your council or look in the telephone directory.

Waste items that are taken for recycling are:
- aluminium — foil and cans, pots and pans, building materials
- batteries — car
- brass
- cardboard — clean waste, clean boxes
- car bodies and car parts
- copper
- containers — steel, empty tanks, 5-200 litre (1-50 gal) plastic, etc
- electric motors
- electric goods

- glass — bottles and jars
- gunmetal
- industrial offcuts — timber, foam, fabric, tiles, etc
- lead — including ballast
- newspapers
- oil — petroleum, vegetable and animal fats
- organic waste — bulk and uncontaminated
- paper — office
- P.E.T. bottles
- plastic — clean bottles, film, bags, milk bottles (labels removed), plant pots
- radiators
- steel — building materials, cans (clean, unwaxed), scrap
- stainless steel
- timber
- zinc

WHAT YOU CAN DO AT HOME

- separate your garbage: recyclables, compost and non-recyclables
- send only those things which can't be recycled to the tip
- use cloth nappies rather than disposable nappies which do not break down
- make a compost heap for vegetable and garden refuse — it reduces waste and nourishes your garden
- use non-disposable items such as ceramic mugs instead of paper or styrofoam cups, cloth instead of paper napkins, and handkerchiefs instead of tissues
- buy second-hand: clothes, toys, books, anything
- don't buy things you don't need. When shopping, be aware of the packaging and the product, and always recycle where possible

Recycling at Home

Lots of bottles, containers, newspapers, and other miscellaneous items can be recycled at home. Here are a few ideas to put these 'throwaways' to good use.

Plastic Containers

- plastic bottles of all sizes make handy funnels — simply

cut out the bottom with a Stanley knife. If you need a fine filter, stretch a piece of clean nylon pantihose over the neck of the bottle

- an automatic 'drinker' for chickens can be made from a 5 litre (1 gal) bottle and a 1¼-2 litre (2-4 pt) cordial bottle.

  Cut a section out of the top of the 5 litre (1 gal) bottle large enough for the cordial bottle to be inserted upside down into it. A small hole is then poked in the neck of the small bottle, about 20 mm (¾″) below the top (the caps are left on both bottles). Next, a slot large enough for the chicken's head is cut into the side of the large container, about 35 mm (1½″)above the bottom.

  Fill the cordial bottle with water and insert it into the 5 litre (1 gal) container, allowing the water level to fill just below the slot.

  This design has two distinct advantages: firstly, the chickens cannot climb into the water and foul it, and secondly, the handle of the large bottle allows it to be tied to the coop to prevent it from being knocked over

- use 2, 4 and 5 litre (4, 8 and 10 pt) bottles as scoops — just cut out the bottom and then part of one side in a

WATERER

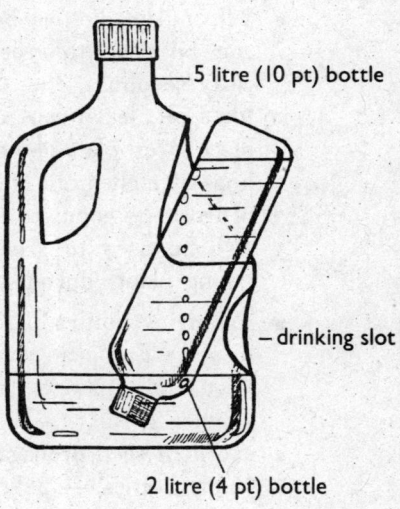

5 litre (10 pt) bottle

— drinking slot

2 litre (4 pt) bottle

STORAGE

semi-circle. They can also be used for the storage of odds and ends, as sorting bins, etc, by cutting out the tops and still leaving the handles intact

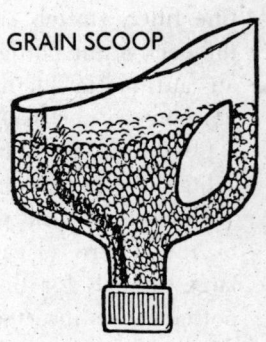

GRAIN SCOOP

- 2 litre (4 pt) bottles, or larger, can be used to greenhouse early seedlings. Cut the bottoms off, leave caps on, and place over planted-out seedlings at night until all danger of frost has gone — they can also be used to greenhouse young plants during the day

- soft drink bottles, 1.25 litre (2 pt) or larger, can be used to make European wasp and fruit fly traps (see chapter 7, Pest and Disease Control) and house fly traps (see chapter 4)

- excellent seed propagating trays can be made from 2 litre (4 pt) bottles — poke a few small holes in the base and then cut off 50 mm (2″) from the bottom. Use the top whenever protection is needed

- roll-on deodorant bottles with removable balls make ideal painting pens for young children. Fill with a non-toxic paint; less mess than paint brushes and small fingers. They also make useful sunscreen applicators, small enough to carry in a handbag, and easy for children to use

- large ice-cream containers can be used as sewing boxes, fishing tackle boxes, or to store shoe cleaning materials and other bits and pieces

- ice-cream containers are ideal for keeping wool and crochet clean and to prevent it from tangling. Cut a hole, about the size of a five cent piece, in the bottom of the container; thread wool or crochet cotton through the hole and replace the lid

- containers and buckets also make ideal planters for growing kitchen herbs. Punch a few holes in the bottom, add a

15 mm (½″) layer of clean gravel and add soil mixture appropriate to the herb's needs

- other ideas for using ice-cream containers:

  cut-out shapes, such as gingerbread men, Christmas trees, Christmas bells, etc, for decorations or templates for biscuits. Decorations can be brightly coloured with paints or felt pens, and templates should be dipped in flour, then placed on biscuit dough and cut around

  animal and map shapes can be cut out and make excellent stencils for school-aged children, and play shapes for toddlers and preschoolers

  oblong shapes with pointed bottoms make good plant markers

- small, flat ice-cream containers make excellent lunch boxes and fit easily into school bags
- round ice-cream tubs are ideal to store play-dough, puzzles, crayons, pencils, and play blocks, and also make useful planter pots
- margarine tubs and take-away food containers can be used for freezing soups, fruit juices, casseroles, etc — three-quarters fill and seal with masking tape, label and date
- use any small plastic containers for seed propagation

SEED PROPAGATION

holes in lid

soil mix

- garden seed packets can be filed in margarine tubs
- yoghurt containers and margarine tubs can be used to store nails, screws, nuts and bolts, and other small odds and ends
- yoghurt containers can also be used to grow individual seedlings, or be used for children's painting pots

Empty Tins and Cans

- large fruit juice cans make excellent bulk storage containers in the home workshop. They can be used as plant pots — paint them a suitable colour and make sure four medium-sized holes have been punched in the bottom to allow air to the roots and water to run through

- brightly painted, large, fruit juice cans make ideal kitchen storage containers and look great on a bench or shelves — use them as a utensil tidy, egg tins, to keep scissors, glue, tape, etc, where you can easily find them, or anything else you can think of

- brightly painted tins of various sizes, with press-down lids, make inexpensive, yet attractive kitchen canisters

- a large tin with the top and bottom cut out, placed around young trees will give support and protection until they have established themselves

- composted tin cans return essential elements to the soil, and especially benefit fruit trees. Simply crush the cans, spread them around the tree and cover with a thick layer of mulch (about 20 cm/8″). After approximately twelve months the can will have completely decomposed, leaving a viable compost.

  You can also compost the tins after crushing by placing them in a shallow hole, then covering with a 15-20 cm (6-8″) layer of dirt. Keep building up alternate layers until approximately 30-45 cm (12-18″) above ground level, finishing off with a good thick layer of mulch over the whole of the mound

- used steel wool can be recycled by burying it in your garden — it will eventually completely decompose, returning essential elements to the soil. Hydrangeas especially love it, turning blue flowers more vivid and pink flowers into an attractive shade of mauve

- a large tin with the top cut out and the bottom punched full of holes will make a watering can when used in conjunction with a bucket of water

- saving aluminium cans will provide extra cash if you can collect enough of them

Glass Jars

- jars with metal lids can be attached to the underside of shelves for the storage of nails, screws, nuts, bolts, washers, etc, which can be easily seen at a glance. Nail or screw lids to the shelf — don't use jars with plastic lids, as they tend to split or break

  The same system can be used to store plant seeds — ideal for serious gardeners who save their own seed

- used wine flagons can be made into miniature cloches to protect young seedlings from frost or cold conditions. To remove the bottom, gently heat the flagon by filling it with hot water; as soon as it is hot, empty water out and stand the flagon in a basin containing about 20-30 mm (¾-1¼″) of ice-cold water. This will cause the glass to break reasonably cleanly at the ice-water mark. A glass cutter will give you a much more precise cut

CLOCHE

- jars of varying sizes are ideal for preserving homemade pickles and jams. Those jars that have the Agee-type twist seal lids can be used for bottling fruit or high-acid tomatoes
- drinking glasses come ready made with some jars — vegemite, cream cheese, honey and nutella. Other jars and bottles can be cut to the appropriate size. File the drinking edge with emery paper before use
- coffee jars make ideal storage cannisters, especially those with air-tight seals
- narrow-necked jars and bottles can be made into decorative lamp bases. For interesting effects fill them with different shells, pebbles or layers of coloured sand

Cardboard Boxes

- shoeboxes and other small boxes with lids, when covered

with wrapping paper or magazine collages, make effective gift boxes

- egg cartons make good seed propagation trays. When seedlings are ready to transplant, cut out individual cups, remove the bottom, and plant out — this eliminates any transplant shock seedlings might otherwise suffer. The cardboard will eventually decompose into the soil
- shoe inner-soles can be made from single-ply cardboard boxes. Use shoes to trace out the approximate size and cut out. Ideal for people who suffer from foot odour — soles can be replaced easily and cheaply
- large boxes can be used as storage drawers, made into cubbyhouses for children or shredded and added to the compost heap

Paper

- windows and mirrors can be cleaned with a wad of newspaper that has been wrung out in vinegar water (see chapter 4). Rub off and shine with sheets of clean newspaper. Used newspaper can then be torn up and added to the compost
- newspaper logs make excellent fuel for combustion heaters or stoves. Roll the newspaper, tightly from the bottom of the page, around a broom handle until it is about 5 cm (2″) thick. Remove the handle, tie the log with string, and then soak it in used cooking oil or fat. When dry, roll more newspaper around it until it is about 10 cm (4″) thick — secure again with string. These paper logs will burn for several hours
- compressed paper blocks can also be used for combustion stove fuel. Paper must first be soaked and pulped, then packed into a suitable mould and compressed.

You can construct a mould from a large fruit juice tin: drill a 50 mm (2″) wide band of 3 mm (⅛″) holes around its base to allow water to escape. Next cut a 'follower' from a piece of hardwood, a fraction smaller than the mould's inside diameter.

Pack the pulp tightly inside the mould, and then place the 'follower' on top of it. Next put the mould under a sturdy shelf or bench, and then position a hydraulic car jack so that its base rests against the underside of the shelf, and the jacking bar against the 'follower'. Operate the jack to compress the paper. Remove from mould and allow to dry thoroughly before use

- brown paper, wallpaper or wrapping paper can be used to cover children's school books
- clean butcher's paper can be used for drawing paper — press with an iron to remove the creases. It can also be used as a cake-tin liner after being brushed with oil or melted butter
- newspapers can be shredded and added to the compost heap
- slugs can be discouraged by placing newspaper barriers in their path. Fold several sheets of newspaper and lay in the rows between young seedlings, dampen slightly. Slugs will attach themselves to the paper and will hide between the folded sheets. Check daily and dispose of slugs. The paper can then be torn and added to the compost. This method is an excellent way of protecting lettuces
- newspapers can be used as the ground level base layer of a new garden bed (see chapter 7 — No-dig Garden)
- cardboard toilet roll cylinders can be used to propagate garden seedlings. Fill cylinders with soil, plant a seed in each one, and stand them upright in an old ice-cream container. When seedlings are large enough, plant them out — toilet roll and all
- labels from canned food can be used as notepaper for shopping lists, phone messages, etc, after which they can be added to your compost
- balls of dampened newspaper are excellent for cleaning dust from dirty flyscreens
- all types of paper can be shredded, pulped and then turned into homemade paper. It has many decorative and other uses and is easy to make

Homemade Paper

Homemade paper can be used to make decorative cards, drawer liners (scented or embellished with fragrant herbs), and for hobby printing, such as lino cuts. I have successfully made varying and interesting paper that I have used for etching and lithography.

You can use old newspapers, computer printout paper, blotting paper, wallpaper, and old telephone books — in fact any paper product can be recycled, including cardboard items like egg cartons, etc.

Before you start you will need the following equipment:

10 litre (2½ gal) bucket, or larger

10 litre (2½ gal), or larger, enamel or stainless steel pan

large plastic basin

2 flat waterproof boards larger than the paper being made

wooden spoon

large wire sieve

rubber gloves

4 G-clamps, or similar

squares of calico

blender (optional)

deckle — wooden frame, A5 or A4 size, with stainless mesh secured to it for limiting the size of the paper and for collecting the pulp. (This can be purchased or made quite easily at home — see instructions this section, A Homemade Deckle.)

Preparing recycled paper

- tear paper into small pieces and put them in the bucket
- cover with warm water and soak overnight
- drain off water and tear wet pulp into smaller pieces, or liquidise in blender — to liquidise, add ¾ litre (1½ pt) of water to 3 tablespoons (4 US tbsp) of pulp and process for 15 seconds; for very fine paper process 25-30 seconds
- put the pulp in a large pan and cover with boiling water
- bring to a rolling boil, and boil for 30 minutes (shredded paper), 10 minutes (liquidised paper), stirring continually with a wooden spoon
- drain pulp off through a large sieve

Making the paper

- place one of the waterproof boards conveniently near to the basin
- lay a piece of calico over the board
- three-quarter fill the basin with pulp and then top with water to just below the rim, mix well

- dip the deckle vertically into the basin, tilt to horizontal below water line and raise slowly, keeping the frame horizontal
- remove top deckle guide frame, tip bottle frame, containing the pulp, upside down and onto the calico-covered board
- press down firmly and evenly to remove excess water, then carefully lift off the deckle
- cover the pulp with another square of calico and repeat the process until all the pulp has been used
- cover the last lot of pulp with calico, place the second waterproof board on top, affix G-clamps to each corner and tighten each one a little at a time to maintain even pressure. Keep tightening until you can no longer apply pressure with the clamps
  NOTE: It is best to carry out this procedure outdoors, since the pressing will expel quite a lot of water
- leave for 2-3 hours, then remove the clamps and carefully lift off each sheet of calico, placing the pressed paper on a flat board to dry. To prevent paper from curling lay another board on top and leave until completely dry

Scented paper   You can add fragrant herbs or flowers or pot-pourri to the pulp once it has been taken up with the deckle, but before it is turned onto the pressing board. Use only dried herbs and flowers.

A few drops of essential oil of your favourite herb may be added to the pulp, in place of the dried herbs, when it is in the deckle.

Alternatively, lay sheets of your paper, once dry, in a sealed box with aromatic herbs scattered between every third or fourth layer. Keep well sealed for at least 6 weeks — this is an easy way to add an elusive air to homemade writing paper. Include envelopes in the same box.

Writing paper   So that homemade paper can be used to write on it has to be treated or 'sized' to receive the ink. This can be done by using store-bought starch or the homemade starch in chapter 3.

Mix $\frac{1}{5}$ teaspoon of starch with a little cold water and mix into the pulp after stage 3 of 'making the paper'. You may have to adjust the amount of starch, depending upon the quantity of pulp being used. As a general guide, the amount specified is sufficient for half a 10 litre (2½ gal) bucket of sodden paper.

A homemade deckle To make the deckle you will require sufficient timber to make two frames as illustrated. Use pieces of timber, such as radiata pine, approximately 25 mm x 6 mm (1″ x ¼″).

The upper frame is constructed on the flat, as with a picture frame, with its inside measurements being equal to the size of the paper you wish to make. At either end, in its centre, a nail is positioned, and its head cut off, to form a locating pin. A corresponding hole is drilled into the top of each end of the bottom frame.

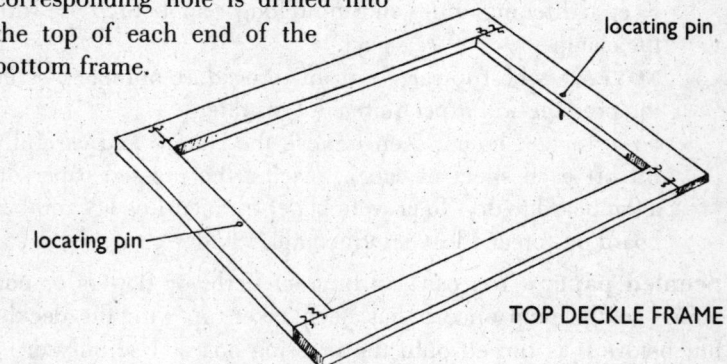

locating pin

locating pin

TOP DECKLE FRAME

The lower frame is constructed on edge, with a supporting rib running across its width approximately every 100 mm (4″). A suitable fine mesh, such as stainless steel, is attached simply by stapling it. Fly-screen mesh can be used instead of stainless steel, but has a tendency to stretch and sag.

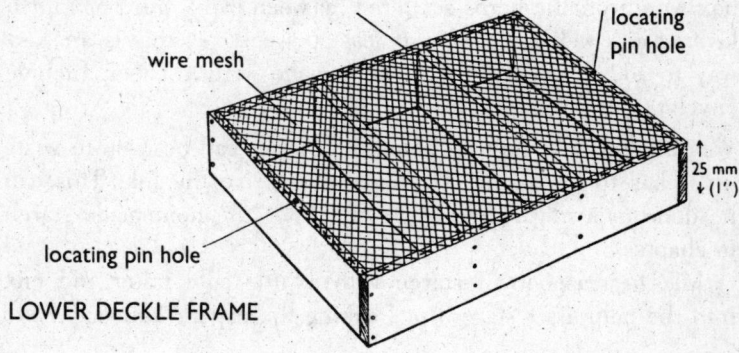

wire mesh

locating pin hole

25 mm (1″)

locating pin hole

LOWER DECKLE FRAME

Milk Cartons

- cut-off cartons make excellent transplant containers for vegetable seedlings. Perforate around the bottom of the carton for easy removal when transplanting time comes
- washed cartons can be used for ice-blocks for portable picnic fridges. Three quarters fill the carton with water and place in the freezer until needed
- use cartons to hold paint when only a little is needed, or to soak paint brushes in

TRANSPLANT   CONTAINER

Plastic Bags

Avoid taking plastic bags from supermarkets and other shops. Use a cardboard box to pack groceries in, or better still, take your own non-disposable bags or a shopping trolley. However, if you do end up with some plastic bags here are some ways of putting them to further use:

- bread bags, and fruit and vegetable bags, can be used as freezer bags. Don't forget to mark each one clearly with item and date
- supermarket, fruit and vegetable bags can also be used to pack school lunches
- shopping bags from the supermarket double as small bin bags and are handy for transporting wet towels and swimming costumes

Food Scraps

- vegetable peelings, tea bags, coffee grounds etc can go on the compost heap
- natural dyes can be made from cold tea and vegetable peelings, such as the skins of onions, pumpkin and beetroot
- cooking oil and fat can be saved and used to make bio-degradable soap (see chapter 3)

Old Car Tyres

There is not a great deal commercial recyclers can do with old car tyres, however discarded tyres can be used around the home, whether you live in the country, have a large garden, or just a small inner city balcony. They will cost you absolutely nothing — tyre dealers are more than happy to give them away:

- try growing potatoes in them. Place a seed potato in the centre of an old car tyre and cover with 150 mm (5 fl oz) of soil to start. As the plant grows add soil to just below tip of shoot, and tyres as required. Usually 5-6 tyres is sufficient. Once the last tyre is full of dirt, allow to flower and harvest. This method will give you a bumper crop and requires very little space. It not only saves space, but makes harvesting easy. Simply brush the dirt away as you remove each tyre, then pick the potatoes

TYRE POTATO GROWER

- cut sections from a tyre, then trace a foot shape onto them using an old thong or shoe. Cut out and use as the sole of a pair of homemade sandals or thongs, or to re-sole old work boots
- cut through and stretched out and laid like 'Roman tiles', tyres make inexpensive and waterproof roofs for garden sheds, chicken shelters, etc
- partially bury into garden soil to control plants, such as the mints, that have invasive root systems
- tyres will also contain mulch around small trees
- stand lightweight garbage bins in tyres to prevent them from being knocked over

- play swings; garden edging; buffers for boats; the use of old tyres is only limited by your imagination

Soap Scraps

Soap scraps can be recycled into new cakes of soap. Place collected soap pieces in a saucepan, cover with boiling water, soak for two days, and then slowly bring to the boil, stirring to ensure that the ingredients are well mixed. Pour mixture into suitable moulds, such as patty pans, and allow to harden.

Rags and Old Clothes

When you think it's time to throw your old clothes in the rag bag, think again! Old clothes, and scraps of material, have many uses. Even if they can't be remade anymore they can still be cut into strips for weaving

- old towelling can be turned into pre-soaped bathroom wash bags. It saves on soap and is easy for small children to handle. Make a drawstring bag, hang in your shower recess and fill with left-over soap pieces
- old garments, or left-over fabric from dressmaking, can be used to make colourful covers for wooden coathangers
- old, hand-knitted jumpers that have lost their shape completely can be unravelled and reknitted. Unravel the wool into skeins, tie, and then wash it. Dye, if you wish, while still tied up in skeins
- unravelled wool can also be used to crochet patchwork rugs, quilts, mats, bedsocks and mittens
- a small jumper can be turned into a tea cosy or an oven mitt. Bind the seam with brightly coloured bias binding or scraps of material. Use left-over lengths of coloured wool to embroider a matching design
- a man's jumper can find new life as a bolero for a smaller person. Remove the sleeves, cut it open down the front, and bind the edges with a brightly printed material
- the sleeves of an old jumper can be made into a pair of warm, cosy slippers
- cut old jumpers into squares, bind the seams, and use as pot-holders, shoe polishing pads, dusters and pieces for a patchwork rug. Left-over scraps can be used to stuff cushions or soft toys

- scraps of cloth can be used to make a patchwork quilt, torn into strips and braided or woven into floor rugs and bathroom mats, made into cushion covers, kitchen aprons, tea cosies, place mats, carry bags for library books or other odds and ends, dolls clothes
- old towels and nappies, when cut up, make ideal polishing cloths. Or sew the best pieces of them together to make beach or picnic rugs — dye them a favourite colour
- left-over bits of rag and clothing, provided that they are natural fibres, can be cut into small pieces and added to the compost heap

Making a Rag Rug The first thing you must do is unpick the discarded garment and tear it into narrow strips about 1-2 cm wide, then sew them together with a diagonal seam. Or you can use braided strips of cloth for your yarn.

Yarn can be woven on a loom, or knitted or crocheted on large needles. When knitting or crocheting double over torn strips, or use braided cloth, to make the rug thicker. For an interesting effect, crochet small squares from different coloured material and then sew them together.

Other useful items can also be made from the woven yarn: bath mats, scatter mats, bedcovers, wall hangings, jackets, handbags and shopping bags, all of which are hard wearing.

Yarn Braiding Tear strips 2-4 cm (¾-1½″) wide from old material or use discarded woollen jumpers or scraps of wool.

Start by stitching three strips together for a 'T Plait', making sure that each strip is a different length. Braid the strips as you would hair, tacking on new strips as required, and keeping seam edges all facing inwards, strips flat, and the work tight.

The length of your yarn will depend upon how you intend to weave it. Rugs and other items made from this type of yarn can be washed and will withstand a lot of rough wear.

Homemade Bag A simple bag made from woven strips of discarded material can be used to carry books, craft work or shopping.

Cut out material according to pattern. Fold material and sew the sides using French seams, then turn inside out and sew again.

Hem the top. Cut two 12 cm (5″) wide strips long enough to make suitable carry straps. Sew them into tubes, then turn right side out and neaten and hem each end and stitch to the bag.

Line with calico, or other suitable cloth, before you sew it together.

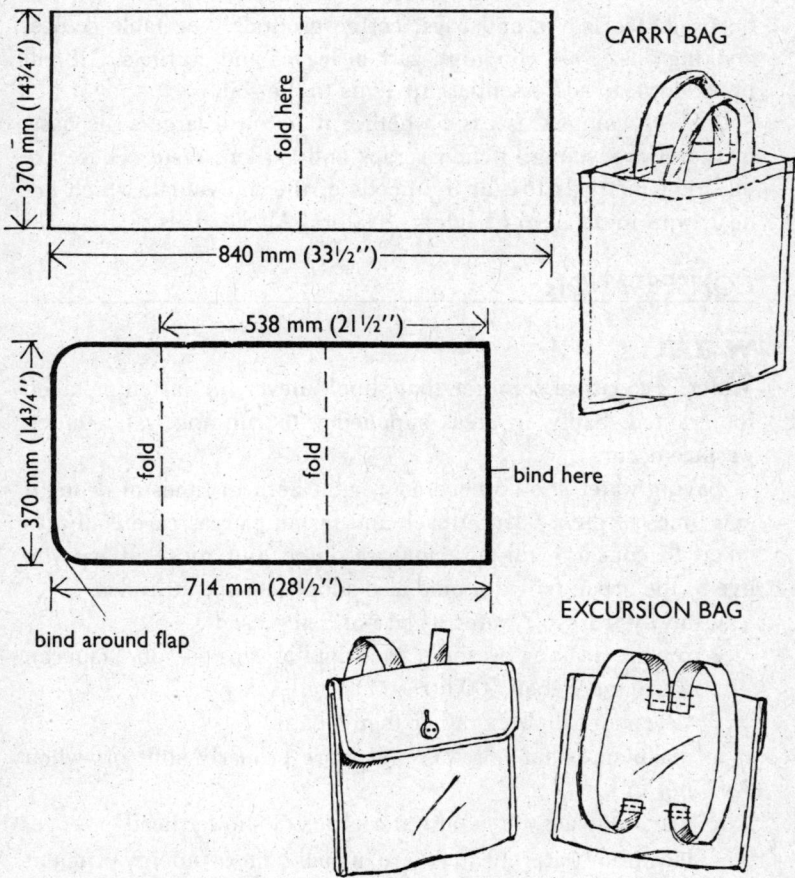

CARRY BAG

370 mm (14¾″)

fold here

840 mm (33½″)

538 mm (21½″)

370 mm (14¾″)

fold

fold

bind here

714 mm (28½″)

bind around flap

EXCURSION BAG

Excursion Bag An excursion bag is made exactly the same as a carry bag, except that it is smaller and has a flap.

Before sewing together, bind the edges with bias binding, make a button hole in the flap and sew a large button on the front of the bag.

COMPOSTING

A large portion of the daily household waste that goes into the garbage bin is food scraps or garden waste. These materials could be put to much better use on our garden as compost, lessening, or even eliminating, the need for chemical fertilisers.

Just about any organic material, excluding meat wastes, can be used: tea leaves and bags, coffee grounds, vegetable scraps, cooking oils, grass clippings, garden leaves and cuttings. All will break down to add essential nutrients to the soil.

Making compost is easy, whether it be on a large scale basis or merely recycling kitchen scraps and garden waste. There are different methods to suit the needs of the individual, which are dealt with in detail in Chapter 7, Natural Alternatives.

CONSERVATION

WATER

Water is a critical resource that should never be wasted or taken for granted. Sadly, in areas supplied with running water, much wastage occurs.

Saving water should become a habit both in times of drought and times of plenty. In outback and farming areas care is always taken to conserve this precious resource, but those of us who live in the urban fringe should also adopt sensible attitudes:

- always turn off water when not being used
- replace leaking washers in dripping taps — one tap can waste more than 500 litres (130 gal) a day
- take short showers rather than a bath
- make sure that hot water taps are properly shut off when not in use
- install tap aerators and water-efficient shower heads
- localised water heaters are usually more energy-efficient than long pipes run to remote bathrooms or laundry
- put an insulation blanket around your hot water system, and wrap insulating material around hot water pipes
- when purchasing a hot water system, choose a solar unit
- install a flow control valve (from the hardware store) at the back of your shower head

- blend hot and cold water when filling a bath
- put the plug in the sink when washing your hands
- don't leave the tap running when cleaning your teeth. Use a cup of water instead
- fill the kettle from the cold tap
- use low flush toilets or put a brick in the cistern — water wastage can be reduced by as much as 40 per cent
- use a bucket when you wash your car
- reduce the amount of water used in your washing machine by keeping to full loads
- use cold water for washing
- when buying a new washing machine, choose one that uses water efficiently
- soak clothes before washing to shift dirt without harsh cleaners
- if you must have a dishwasher, keep to full loads — and when buying one, find out how much water it uses. However, old fashioned elbow-grease will clean just as well and conserve water
- install a rainwater tank to make use of roof run-off. It can provide great drinking water, unadulterated by chemical additives, can be used for an efficient garden drip watering system, and connected to your laundry it will reduce demands upon supplied water. (With most water supply authorities now charging for excess water use, a supplementary rainwater tank makes sense)

To cut down on garden watering

- mulch the soil surface around trees and shrubs and on garden beds — it can prevent up to 73 per cent evaporation loss
- water the roots, not the leaves of the plants. This encourages deep root growth and makes plants hardier
- if you have a large number of ferns or annuals, use a micro-spray; the water is distributed at a rate the soil can absorb, reducing run-off
- consider sowing tougher grasses for lawn that aren't so water dependent — kikuyu, couch, Kentucky bluegrass and perennial rye

- plant natives — they are hardy and need less water
- never water after dark; it adds to the risk of forgetting to turn taps off
- soaker hoses are an efficient way to water lawns and garden beds, providing a light mist of water which soaks in very effectively, especially during very hot, dry or windy weather. They are also excellent for running through vegetable gardens, where plants need watering at ground level only
- drip watering systems, which consist of a length of black plastic tubing with tiny holes to release very slow drips of water at the base of plants, are ideal for country gardens where water is limited
- 'grey water' (water from baths and washing machines) can be siphoned off into a tank for watering vegetables and ornamentals, provided you use pure soap or biodegradable washing products

ENERGY

Today, almost everyone is feeling the energy pinch; the cost of utilities is skyrocketing, and is increasingly affecting our lives and livelihoods.

We can as individuals take an active role in helping to alleviate the problem. By re-evaluating our needs and using energy wisely we can help to conserve community energy sources. This is especially so in the home, where the inefficient use of energy is often the greatest.

Electricity

- minimise your use of lights and electrical appliances
- switch off lights that are not essential and use energy-saving lightbulbs
- install dimmer switches — they save energy
- boil only what water you need in an electric kettle
- check that your electrical appliances operate efficiently. Well maintained and effective door seals on refrigerators and freezers make a big difference. Once they deteriorate, replace them
- put on more clothes before you put on a heater. Avoid excessive use of heaters as they have a high power

consumption — use them when only absolutely necessary

- don't waste hot water; have shorter showers in winter and cold showers in summer
- open refrigerator and freezer doors as little as necessary, and then close again as quickly as possible. Opening the door frequently uses a lot more electricity
- minimise the purchase of consumer goods — their manufacture consumes a considerable amount of power
- when purchasing goods, support those manufacturers who use energy-efficient methods
- check the energy consumption of products before you buy. Fridges and freezers have a star rating system to indicate the amount of electricity they use
- adjust thermostats to settings recommended by the manufacturer of the appliance
- make sure your refrigerator and freezer are well treated with sufficient space around them so that the heat can escape
- keep condenser coils at back of refrigerators and freezers clean and free of dust and lint
- opt for the smallest fridge possible
- consider sharing a freezer with a neighbour. Freezers run more efficiently when full
- don't buy electrical appliances if there is a manual one that will do the job
- use the clothes line rather than a clothes dryer
- put an extra blanket on the bed instead of an electric blanket
- defrost refrigerators and freezers when required; space foods to allow air circulation; do not overload; do not put hot foods in the cabinet; cover liquids and solids before storing
- use natural gas instead of electricity for hot water, heating and cooking
- when you go on holidays, turn off heating, electrical appliances and the hot water system
- if possible, repair faulty appliances rather than throwing them away
- if building your house (or renovating) incorporate passive solar design for cooling and heating — see chapter 2

Heating and Cooling

- on hot days keep windows and external doors closed, lower external blinds or awnings, and draw internal blinds or curtains. Close internal doors which lead to areas likely to be influenced by outside conditions
- instead of using an air-conditioner in hot weather, open all the windows and let the natural breeze cool you. Use simple, effective and inexpensive solar ways (as detailed in chapter 2) to cool your house
- in winter, draw the curtains to help retain warmth
- wear warmer clothes rather than turn up heating
- close off unused rooms in winter and do not heat rooms that are not being used
- make sure you keep your home well insulated throughout
- fit draught excluders on all external doors
- fit draught excluding tape around door jambs and window frames
- chapter 2 deals far more extensively with energy-efficient means of heating and cooling the home

Cooking

Top of the Stove

- cook vegetables in the least possible amount of water. Once something has boiled, turn down to low heat and simmer
- cook vegetables only until tender, and with the lid on the pot
- don't repeatedly lift the lids of your saucepans
- try to make 'one pot' meals, cooking different vegetables in a large pot with dividers; or use stackable sets of twin or triple saucepans
- aluminium foil under electric hotplates will reflect the heat upwards — keep clean for maximum efficiency
- use the correct sized pot — don't waste energy with small pans on large hotplates
- a crockpot or frypan is cheaper to use than a conventional cooktop for the same meal
- on electric stoves, turn off the element a few minutes before the food is fully cooked
- use good quality utensils made from heavy cast iron, enamel

or stainless steel and with thick bases (such as copper) — they heat up far more quickly, and retain heat, spreading it evenly so that less energy is used to cook the food

- cook efficiently — use a pressure cooker or steamer to cook several things in the same pot
- do not use the grill for toast; a toaster uses less energy
- eat more fresh and raw food

In the Oven

- roasting and baking are most energy consuming. However if cheaper cuts of meat are marinated first they will require less time to cook
- plan ahead, try to cook more than one dish at a time. Double the quantity and freeze it for later, or bake a loaf of bread as you cook a casserole. Most importantly, try and use all the racks
- avoid opening the door. About 20 per cent of the heat escapes each time
- when baking bread use dark or blackened bread pans
- cook food in ceramic, ovenproof glass or cast iron casseroles, heat is transferred more efficiently and retained for longer
- turn off the oven 10 minutes before the allotted cooking time. The food will continue to cook and you'll save that much more energy. Set the stove timer to remind you
- don't buy a stove with an oven bigger than you need
- whenever possible, use a microwave oven — it will reduce cooking costs by up to 30 per cent

Fireless Cooking

I was first introduced to this type of cooking when I was a member of the Boy Scouts. It consisted of a device known as a 'hay box' in which you could cook casserole and soup type dishes.

A wooden box, and its lid, was lined with several thicknesses of newspaper (each layer glued on top of the other) and the box then filled with uncut hay. The meal was quickly brought to the boil in a 'billy can' and then buried inside the hay. More hay — stuffed inside a hessian bag to form a pillow — was pushed down on top and the lid held in place with a heavy stone.

The meal was prepared in the morning and left to cook, using its own heat, during the day while we were away enjoying an activity. When we returned in the evening, tired and hungry, there was a hot meal waiting for us.

A modern version, the hot-box, is simple and easy to construct. It is ideal for cooking casseroles, soups, porridge, rice and rice dishes, cracked wheat and other grains, dried beans, and cheaper cuts of meat that require slow and even cooking. You can also use it to keep cooked food warm and to make yoghurt. With double, or triple stacking pots you can cook a complete meal.

Construction of Fireless Cooker The box is made from 12 mm (½″) thick plywood, its size depending upon the pot you will be using.

Construct a square box 200 mm (8″) larger than the outside diameter of your pot. It should be 100 mm (4″) taller than the pot with its lid in place. Use 25 mm (1″) brads and wood glue to assemble the box.

HOT-BOX

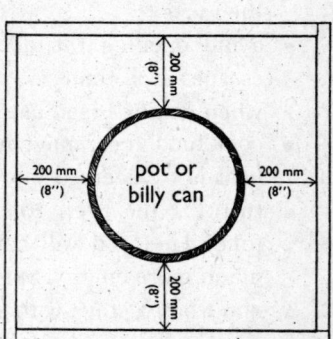

To make the lid, start with a square of plywood with dimensions 40 mm (1½″) greater than the outside dimensions of the box. Cut dressed 100 x 25 mm (4″ x 1″) pine (this will dress down to 90 x 19 mm/3⅝″ x ¾″) to fit the edges, and nail and glue the pieces in place to make a tight fit.

You will need a styrofoam block, or sheets of styrofoam that you can glue together to form a block large enough to fit snugly inside the box and the lid space. Cut off the section that fits inside the lid, glue in place, and remove foam, if required, to form a cavity for the pot lid. At the most, this should only be a shallow indentation.

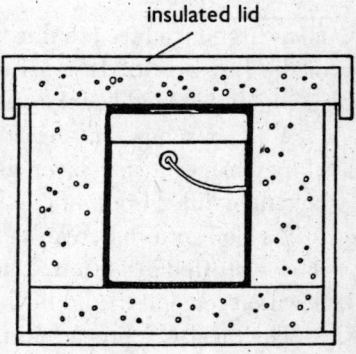

insulated lid

Cover the foam with aluminium foil, shiny side up; fasten with glue.

The remaining section of the block is now hollowed out to the shape of the pot. This can be done easily if the block is first cut into quarters, the appropriate section of the pot profile removed and the sections glued back together. Glue aluminium foil, shiny side up, to the top and insides of the block, and then glue it in place inside the box.

Use two 25 mm (1") butting hinges to attach the lid. Install a light hasp to the front edge of the lid and a staple to the front of box.

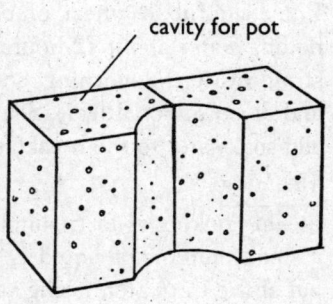

cavity for pot

Cooking with the Hot-box   Food is cooked in the pot on your stove until it boils, and is then transferred with the lid sealed to the hot-box. The hot-box is closed and the food left for about three times as long to complete cooking. It is important that food is boiled to kill bacteria before being transferred to the hot-box.

Food cooked by this method won't overcook, boil over, burn or dry up.

Dried beans should be soaked overnight before cooking, and porridge for breakfast should be put in the hot-box overnight. Rice, oatmeal, and other grains cooked this way will swell to the maximum.

Benefits of a Vacuum Flask

A wide-mouthed vacuum flask is a great energy saver. It will keep food hot for hours, save hot water, soak dried grains, vegetables and legumes, and can be used to cook in by simply adding hot water.

Rice and Vegetable Stew
Add diced vegetables, rice, lentils and stock to a saucepan. Bring to the boil, allow to boil for about two minutes, and then pour into your vacuum flask. Cap and leave for 3-4 hours to cook.

Winter Breakfast

Put a cupful of wheat or oats in a flask, add 1 litre (2 pt) of boiling water about 12 hours before the time you want breakfast, strain off in the morning, and serve with honey, sultanas, raisins, and diced seasonal fruit. Soak raisins and sultanas overnight, and add soakwater to the breakfast if you wish.

Steaming

Steam cooking is fast, simple and energy efficient because only a small amount of liquid is needed. Food is cooked in a closed pot above the water, losing very little flavour and nutrition.

Vegetables, fish, chicken pieces, puddings, and top-of-the-stove breads are delicious cooked this way.

White Wine Steamed Chicken

1 chicken, cut up	*3 slices fresh ginger*
2 tbsp (2²/₃ US tbsp) white wine	*2 tsp salt*
	½ cup water
1 leek, cut in small pieces	

Place the chicken pieces in the steamer, add the leek and ginger, and sprinkle with salt, wine and water. Place the steamer in a pot filled with enough water to steam for an hour.

This recipe is as equally delicious when made with fish fillets. Experiment with cooking times.

Pressure Cookers

Pressure cookers save both time and energy, quickly tenderising food and retaining its flavour and nutritional value, especially vitamin C. It usually takes only one-third of the conventional cooking time for most dishes, and is safe and simple to use.

To get the most out of your pressure cooker, and to ensure that it remains in perfect working order, always follow the manufacturer's directions. When buying, always select a heavy duty unit; it will last and usually comes with a long guarantee.

Drying Food — Energy Efficient

With the availability of canned and frozen foods, drying foods has almost become an obsolete means of preservation. Drying, however, is an excellent natural method by which the home gardener can preserve fruit and vegetables. Nothing is added and only water is removed. With the exception of some loss of vitamins A and C, both fruit and vegetables contain the same vitamins and minerals as fresh produce.

Dried food conserves storage space and is energy-efficient — unlike the freezer unit, it is not using electricity until it is eaten.

Fruit and vegetables may be dried using an energy-efficient dryer or a solar dryer. The most important thing is to remove 80 to 90 per cent of the water content, so that spoilage bacteria can't develop during storage. Suitable storage containers are tightly sealed glass jars.

Fruit and vegetables suitable for drying are:

apples	*beans*
apricots	*cabbage*
berries	*carrots*
cherries	*corn*
coconut	*mushrooms*
dates	*onions*
figs	*peas*
peaches	*peppers*
pears	*spinach*
plums	*squash (all types)*
prunes	*tomatoes*
rhubarb	

Potatoes and most other root vegetables will store well under cellar type conditions, so should be stored whole instead of dried.

Select only blemish-free vegetables and fruit of the highest quality and at peak of maturity. Small fruits like cherries, plums and strawberries may be dried whole, while larger fruits, such as apples and pears, should be cut into uniform sized slices or wedges so that they complete the drying process at the same time. For maximum nutrition, leave the vitamin and mineral-laden skins on the fruit.

Low-Cost Dryer   The materials required for this unit are very simple: a topless cardboard fruit box at least 200 mm (8″) deep; a light socket, base, and cord; a 60-watt bulb; a metal biscuit tray to fit the box; some aluminium foil; a block of wood 100 mm (4″) square; and a small tin of black paint. You can construct the box from scrap timber.

Start by notching one top corner of the box so the electrical cord can exit; then line the box with aluminium foil, shiny side up. Cut the block of wood in half on the diagonal, and mount the light socket on the diagonal surface. This will ensure that the bulb is at 45 degrees when placed in the dryer, and will help diffuse heat more evenly. Position your lightbulb setup securely in the centre of the box, run the cord out through the notch, and place a little foil on top of the bulb (this will also aid in even heat distribution).

Paint the bottom of the biscuit tray with black paint; this ensures maximum heat absorption. When dry, place the tray over the box (black side down) so that it is suspended a few centimetres above the lightbulb.

Fill the tray with a layer of sliced fruit or vegetables, and plug in. Usually it takes about twelve hours to dry the food; depending upon whether it's a very humid day and upon the size of food slices.

The cost of running the unit, based on 1991 charges by the Sydney County Council of 9.52 cents per-kilowatt-hour, is 7 cents to process a batch of apples in twelve hours. This compares very favourably with an electric oven, which, at 100°C (212°F) with the door slightly ajar, and taking about four-and-a-half hours, would cost 42 cents.

Solar Dryer   This works on the same principle as the low-cost drying box, except that the heat source comes from the sun.

Fruit and vegetables should be spread out on trays, made from a wooden frame and covered with muslin cloth. These trays are then stacked in the cabinet with an 8 cm (3¼″) gap between each one for air circulation.

Remove food each evening until dry to avoid moisture re-occurring.

Preparing Fruit For Drying

Apples	pare, core and cut into thin slices or rings. Don't peel unless the apples have been heavily sprayed
Apricots	cut in half, remove seed, and leave in halves or cut into slices or pieces

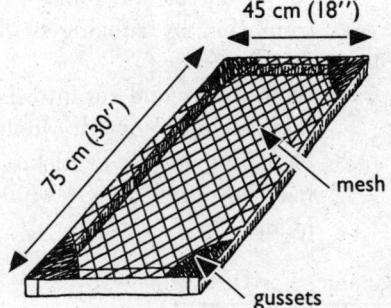

DRYING TRAY

45 cm (18")

75 cm (30")

mesh

gussets

Specifications
cavity of walls lined
with insulation material
inside of door lined with
10 mm (²/₅") rigid foam
absorber surface lined with
10 mm (²/₅") rigid foam

remove plants each night,
replacing again in the
morning

hot air out
(controlled by
adjustable
flap)

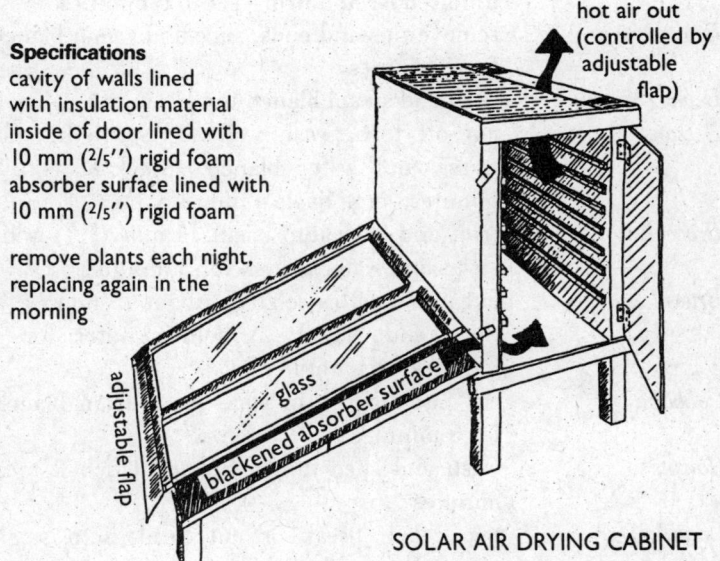

adjustable flap

glass

blackened absorber surface

SOLAR AIR DRYING CABINET

Bananas	peel and slice thinly
Berries	halve strawberries, leave smaller berries whole. Crack skins by quick blanching or nicking with a knife
Cherries	pit and remove stems and drain until no more juice flows from them
Grapes	remove stems and crack skins by blanching quickly or nicking with a knife. Drain until no more juice flows

Peaches	cut in halves and remove pits. Remove fuzz from skin by rubbing briskly with a towel, then slice
Pears	remove core and cut into slices or rings
Plums	may be pitted or left whole. Crack skins by blanching quickly or nicking with a knife
Prunes	may be pitted or left whole. Prepare as for plums

Preparing Vegetables For Drying

Asparagus	use top 75 mm (3") of spear. Steam blanch until tender and firm — 4 to 6 minutes
Beans (runner)	remove tips and ends, slice and steam blanch for 2 minutes
Beans (broad)	shell and steam blanch for 1¾ minutes
Beetroot	cut off tops, wash well and scrub in cold water, and water blanch. Small beets 2½ minutes; large beets 4 minutes
Broccoli	trim and slice into small 15 mm (½") wide strips and water blanch for 3 minutes
Brussels sprouts	cut into lengthwise strips about 15 mm (½") thick and blanch in boiling water for 3 minutes. Dry until crisp
Cabbage	cut into long, thin slices and steam blanch for 5 minutes
Carrots	wash and slice thinly. Steam blanch 2 to 3 minutes
Cauliflower	trim, then break or cut heads into small pieces. Steam blanch for 4 minutes
Corn	husk and remove the silk, then water blanch the whole cob for 10 minutes to set the milk. Remove kernels before drying, making sure not to include any of the husk
Egg plant	peel and slice into thin strips or dice into small cubes. Place in a salt solution (1 tablespoon (1⅓ US tbsp) salt to 1 litre (2 pt) cold water) to prevent discoloration before blanching. Water blanch for 4 minutes

Garlic	see onions
Leeks	see onions
Marrow	see zucchini
Mushrooms	peel and cut off stems if they are tough. Leave whole or slice, depending upon their size. Do not blanch, but dry raw
Onions	peel and slice, or peel and grate. Onions and leeks can be cut into rings. Water blanch onions and leeks for 1½ minutes if you intend to use them in soups or casseroles. Do not blanch onions, or garlic and leeks, if being used for seasoning
Parsnips	cut off tips, peel, wash and dice into small cubes. Steam blanch for 1½ minutes
Peas	shell and select only young, tender peas. Water blanch for 1 minute
Peppers, sweet	wash and remove stem and seeds. Cut into thin slices and water blanch for 2 minutes
Peppers, hot	if possible, do not pick until they are mature and fully red. Although dried hot pepper can be stored in glass containers, they are best left hanging in strings in a dry place. String the peppers by running a needle and thread through the thickest part of the stem. Grind into powder for cayenne pepper
Potatoes	peel, wash and slice into 6 mm (¼") rounds. Steam blanch for 5 minutes and then soak in ½ cup lemon juice mixed with 2 litres (4 pt) of cold water for about 45 minutes to prevent oxidation during drying
Pumpkin	clean and cut into 25 mm (1") strips and then peel. Steam blanch for 3 minutes
Rhubarb	cut into strips (about 20 mm/¾" wide) and water blanch for 3 minutes
Spinach	cut very coarsely into strips and steam blanch for 3½ minutes. Spread no more than 12 mm (½") thick on drying trays
Squash, summer	see zucchini

Squash, winter	see pumpkin
Swiss chard	as for spinach
Tomatoes	wash, quarter and water blanch for about 5 minutes. Run through a food mill to remove skins and seeds, and then strain pulp through several thicknesses of cheesecloth. Squeeze out as much water as possible, then spread the pulp out on the metal biscuit tray (if using a solar dryer, the metal tray can be placed on top of the drying trays). Turn the drying pulp frequently until it becomes dry flakes
Turnip	wash and cut into thin slices or cubes. Steam blanch for 1½ minutes

Processing Produce

Fruit is sufficiently dehydrated when it becomes leathery — dry and shrivelled on the outside and only slightly soft inside. Beans, peas and corn should be very hard; leafy and thin vegetables should be brittle; and larger vegetable chunks and slices should be leathery. If at all in doubt, leave the produce to dry a little longer.

Some pieces of food will dry faster than others, so it is important to remove pieces as they dry rather than wait until every piece is completely dehydrated. Fruit appears to be moist when it is hot, so occasionally remove a few pieces from the dryer and allow them to cool before determining if they are dry.

Pasteurising Fruits and Vegetables

Pasteurising is important to guarantee proper storage of dried fruits and vegetables, and will ensure that no insect eggs or harmful spoilage organisms will develop. Low heat will have dried the food, but may not have killed all such contaminants.

Spread dried food 25 mm (1″) thick on metal biscuit trays and heat for 10 to 15 minutes in an 80°C (175°F) oven. Cool thoroughly before storing.

Rehydrating

To rehydrate dried produce, add three cups of water for each cup of dried food. Allow to soak until swollen and the liquid is returned to the food.

Vegetables, except for dried beans and peas, will usually absorb all the water they are capable of retaining in about two hours. Fruits require a longer soaking time — anywhere from two to several hours. If the water is absorbed too quickly, add more — a little at a time — until the food will hold no more. Avoid adding too much water, since nutrients will be lost to this extra water.

AROUND THE HOME

Learn to use safe products in and around the home, and adopt habits that are not harmful towards the environment.

- resist the urge to over-clean
- resist the use of harmful chemical cleaning compounds
- use biodegradable, phosphate-free soap powder for washing clothes
- handwash delicate garments with pure soap
- avoid 'dry-clean' only garments — dry-cleaning solvents are a hazard. If you have to dry-clean, do it at home with a natural, safe product
- deodorise and spot clean carpets using the natural and friendly way
- dry clothes outdoors — the sun is an excellent sanitiser and does not waste energy
- do less ironing — folding clothes neatly as you take them off the line helps to avoid wrinkles
- don't use chemical-based oven cleaners
- use natural, safe plant-based alternatives to control indoor insect pests. Don't use pesticide bombs or similar products, they usually contain dangerous chemicals which linger
- use safe, plant-based air fresheners to help clear the air of noxious fumes or stale odours. Indoor plants will also help to freshen the air
- use safe 'herbal' sprays for the control of garden pests and disease, and organic methods for the fertilisation of plants
- switch to natural cosmetics and toiletries
- avoid animal and petroleum based products — instead, change to natural alternatives based on vegetable extracts
- choose toothpastes without such additives as sweeteners and

colourings. Better still, make your own from natural ingredients
- don't use toxic deodorisers or water colourers in the toilet
- avoid the use of chemical insect repellents. Safe, natural alternatives are best.

Recipes for easy to make natural products that will do everything from clean your house to clean your teeth, can be found under their appropriate heading in part two of this book: 'HERBS AND OTHER NATURAL ALTERNATIVES'.

Avoid CFCs
- boycott aerosol spray cans
- use pump-spray containers for insect sprays, etc
- don't buy take-away foods in foam plastic containers. Make a habit of taking your own re-usable container to be filled with take-away food
- avoid buying products that are packaged in foam polystyrene
- don't dump your old fridge at the tip. Ask the manufacturer, or local repairer, to take the cooling unit and save the CFCs

Friendly Habits
- don't tip unused paint down the sink. Both paint and turps should be wrapped securely and placed in the garbage bin
- use water-based, not oil-based paint. Better still, make your own from biodegradable ingredients — see 'Household Paint', Household Products and Skills, chapter 4
- don't clean paint brushes over a drain. Wash them out in a bucket and dispose of the waste in a hole in the garden away from plants
- never dispose of insecticides, kerosene, sump oil and other toxic substances by burying them or pouring down the drain. Contact your local council or environment authority for disposal advice
- pesticides, chemical fertilisers, and weedkillers should be handed in to your local council for disposal. So should any other dangerous substances — never dispose of them yourself
- save left-over cooking oil for homemade soap. Strain to remove food particles, which can be added to the compost — see chapter 3

- wipe oil or fat residue off dishes or utensils with a sheet of newspaper before washing — paper can then be added to the compost heap
- cut back on use of detergents and use only those which contain no phosphates and are biodegradable. See chapter 3 for homemade, safe alternatives
- don't put milk, tea leaves or coffee grounds down the sink. Tea leaves and coffee grounds go in the compost, and milk can be recycled into a water-based paint — see Household Paint, chapter 4
- fit a strainer in the drain hole of your sink. This will catch any minute food scraps which can be then added to your compost. Scrape dinner plates, and other utensils, as best you can to remove left-overs before washing them
- keep a bucket, with a tight-fitting lid, handy to the kitchen. To this add your scraps for the compost — empty regularly and rinse out. Deodorise occasionally with bicarbonate of soda to eliminate any smell. Rinsing out with lavendar water, and then leaving a little of the rinse water in the bottom of the bucket will leave it smelling fresh and clean. Left-over lavender water can be stored in a tightly-sealed bottle in the refrigerator. Instructions for making lavender water are in chapter 8
- use unbleached tissues and toilet paper — dyes on coloured paper pollute as does bleaching
- condoms, tampons and pads should never be flushed down the lavatory — wrap them securely and then dispose of them in the garbage
- don't use an incinerator to dispose of waste — burning plastics and glossy paper gives off poisonous gases.
- avoid burning leaves, grass, twigs, etc; compost them instead
- try and become more self-sufficient and reduce your dependency on processed foods
- set an example for others in the fight against pollution

HOME IMPROVEMENTS

- whenever possible, choose building products that are biodegradable, use less energy in their production, and do not pollute the environment or add to the depletion of natural resources

- do not use native rainforest timbers (unless recycled) or imported tropical hardwoods. Choose plantation grown timbers, such as Cypress and Hoop Pine, or our beautiful and varied eucalypts and acacias
- avoid mouldings, doors, flooring and furniture that have been made from imported hardwoods
- for outdoor furniture or pergolas refuse those timbers treated with dangerous chemicals, such as pentachlorophenol. Instead, use eucalyptus, western red cedar, and Douglas fir (Oregon pine)
- use natural wood finishes like beeswax and linseed oil
- try and avoid using particle board or plywood — both these timbers contain formaldehyde, which can irritate the eyes and cause headaches, rashes and respiratory problems
- when buying carpet and underlay, or other floor coverings, choose natural fibres instead of synthetics — they use less energy to manufacture

IN THE GARDEN

- grow your own organic fruit and vegetables. This not only guarantees fresh, chemical-free produce, but reduces the need to transport it
- even if your garden is small, or you don't have one, vegetables and fruit trees can be grown in tubs, pots or boxes. Fruit trees can be trained to grow flat against a wall, or a fence, as fans, cordons or espaliers. This not only saves space, but in some cases can considerably increase yields. Citrus trees will readily dwarf when grown in large tubs, while still producing ample fruit. Soft fruits such as blackberries, raspberries, currants, chinese gooseberries (Kiwi fruit) and grapes can be grown against a trellis; as can passionfruit, bitter pear *(Momordica charantia)*, rockmelon, butternut pumpkin, peas and beans. Strawberries can be grown in large tubs or drums, with holes cut into the sides and the plants poked in
- herbs can be grown in window boxes, and in small pots

in the house as long as they receive sufficient sunlight
- attract birds to your garden with bird-baths and flowering and fruiting trees
- encourage children to have their own garden patch — both flowers and vegetables. Radishes are a quick and easy-to-grow vegetable for children to start with
- use only organic control methods for garden pests and disease, and other natural alternatives to maintain a healthy garden environment
- make a compost heap to turn your food and vegetable wastes into a soil conditioner
- share powered garden tools, such as lawnmowers and whipper-snippers, with a neighbour. A far better, energy efficient alternative is to use a push-mower and hand-operated edge cutters

ENERGY-EFFICIENT CLOTHING

- buy 'seconds' or faulty clothing direct from the manufacturer. Sometimes you'll find it on sale at markets. Usually, you'll have a hard time finding the fault
- buy second-hand clothing at opportunity shops or flea markets
- if clothing still has plenty of life left in it, but no longer fits, pass it onto someone else or give it to clothing outlets such as the Red Cross, the Salvation Army or St Vincent de Paul. Those that are beyond hope can be recycled as suggested in Rags and Old Clothes, this chapter
- don't pass up jumble sales, garage sales, school fetes, trash-and-treasure markets, or second-hand clothes advertised in your local newspaper's classified section
- revamp clothes to the current style — make two garments out of one bought cheaply
- If you're worried about the latest fashion, have jackets, suits and casual coats or pants re-cut and tailored. It's still cheaper than buying new clothes and will eliminate unnecessary wastage. If tailoring costs are a problem, enrol in a TAFE course and learn the basic skills yourself. You may even find that this skill can become a source of income

by altering clothes for others, or re-tailoring bought second-hand clothes and then selling them

- buy patterns and make your own clothes — form a club to share the cost of patterns and the exchange of used materials and clothes
- keep a rag-bag for all those clothes that are no longer possible to salvage — they still may have a useful life — see Rags and Old Clothes, this chapter
- old nylon stockings or pantyhose have many uses: storing onions and garlic bulbs (tie a knot in the hose between each bulb); straining liquids; covering a glass jar for sprouting; training climbing plants
- repair your own shoes — even the soles and heels of synthetic runners can be given new life. Use one old pair to resole and heel another pair.

SHOPPING FOR A BETTER PLANET

When out shopping it is important to select those products which are environmentally friendly. Choose goods with the least packaging; that are recycled or recyclable; repairable, refillable, re-usable and reliable. Most importantly, avoid disposables.

The following hints will help you shop for a better planet:

- take your own bags or basket shopping and refuse plastic shopping bags
- cut packaging — buy in bulk. It's also cheaper and reduces the use of your car
- join a local food co-operative or start one with a group of friends for even greater savings on bulk buying
- choose items that have the least packaging, and whenever possible packaged in recycled materials
- buy products which are environmentally conscious, and refillable or re-usable
- avoid buying plastic products, and refuse to use disposable items such as plastic cups, plates, utensils or razors
- buy greaseproof paper as an effective non-polluting alternative to cling plastics
- buy unbleached products and refuse foods with synthetic additives

- read labels — don't buy goods or food which have been heavily processed
- don't buy aerosols — hydrocarbons add to the greenhouse effect. Choose pump action sprays and refills instead, or go without
- buy goods made or grown locally — this reduces the need to ship goods over large distances which otherwise uses energy and helps to cause pollution
- buy organically grown food as much as possible — it doesn't add to pesticide levels
- do not buy endangered plants, or products made from overexploited species
- avoid buying furniture that is made from plastic or tropical hardwoods
- buy non-disposable items, such as ceramic mugs, cloth instead of paper napkins, and handkerchiefs instead of tissues
- buy cloth nappies rather than disposable ones, which do not break down
- buy pure soap rather than detergents; it is phosphate-free and biodegradable. Or make your own pure, biodegradable soaps and washing liquids — see chapter 3
- buy and use rechargeable batteries
- recycle resources by buying second-hand goods

OUT AND ABOUT

Here are some positive attitudes and sensible habits to adopt when using a car to conserve fuel and help reduce harmful emissions:
- keep the engine tuned, to ensure efficient use of fuel
- make sure all your spark plugs are firing — clean them on a regular basis
- drive smoothly, avoiding frequent hard braking or sudden acceleration
- always keep tyres inflated to the correct pressure. Under-inflated tyres will increase petrol consumption
- don't keep your foot on the brake when driving
- remove roof-racks when not in use, and unwanted luggage from the car

- don't overload the boot and don't overfill the petrol tank (remove the fuel hose when it first clicks off)
- switch off your engine in a traffic jam
- ask your local garage how you can effectively recycle and re-use motor oil, tyres and batteries
- choose unleaded petrol — most cars manufactured after 1980 will run on it
- shop locally and walk there
- avoid using the car for short trips; walk instead
- organise a car pool in your neighbourhood for shopping or going to work, and share the expenses
- use public transport as often as possible, especially when going to work
- ride a push-bike to the shops, station or even to work
- if buying a new car, select one that is lightweight and has a low fuel consumption
- avoid using the car air-conditioner because it causes the engine to burn more petrol. Don't buy a car that is fitted with air-conditioning. On hot days it is far more energy efficient to open the car windows and allow the breeze to cool you. For cold weather have a heater fitted that runs off the engine cooling system; it's just as efficient and is pollution-free

Think before you use your car. If you can walk, or use public transport, do it. You will not only be eliminating unnecessary fuel usage, but helping to minimise pollution.

Tree Planting

- reaforestation and backyard and community tree planting will help. However, it is only a temporary solution; sooner or later, when the trees die and rot away, the carbon dioxide is released back into the atmosphere
- planting trees is a positive step, not only for the environment but for a greener planet

IN THE COUNTRY

Involve your family and friends in regular trips to the countryside. This will help them to develop an appreciation of our environment. And of course you don't need to drive the car; consider group cycling, or public transport and then trekking.

When out and about always have absolute respect for the flora and fauna:

- don't litter
- educate your children and others
- admire the beauty and magnificance of what's around you, but don't destroy it. Stay on the path when you are out bushwalking

COMMUNITY INVOLVEMENT

There are numerous ways in which you can become involved in environmental issues. Do whatever feels most comfortable for you and your family — from implementing the responsible habits already discussed to joining environmental organisations. Whatever you do, it is important to start doing something now!

A few ideas for helping are:

- become informed about environmental issues
- express your environmental concerns to councillors and state and federal politicians
- join with neighbours, and other conservation groups, who share your concerns
- organise tree planting days with family and friends
- endeavour to create community awareness of environmental issues: local meetings, posters, letters to local newspapers, etc
- organise clean-up campaigns, and report any illegal rubbish dumping in your area
- adopt an area, such as a park, beach or bushland, to look after as a family or neighbourhood project
- if not already implemented, ask your local council to organise regular recycling pickups
- approach local schools, supermarkets and shopping centres about setting up a recycling scheme
- petition your council not to spray weeds along your street or near schools or children's playgrounds — instead, organise neighbourhood and school groups to do the weeding by hand.

ENERGY-EFFICIENT HOMES 2

SAVING ENERGY

Adequate insulation is the first step to an energy-efficient home. It helps to maximise the effectiveness of heating and cooling systems, by trapping winter warmth and excluding summer heat.

However, before you start to insulate you must first eliminate any potential energy leaks. These usually occur around and under doors, window sashes and panes, floor boards and under badly fitted skirting boards. Not only does cold air find its way in, but warm air readily escapes. Install draught excluders at the base of doors and foam tape around door jambs. Windows should have effective storm mouldings and panes must be sealed tight with putty. And skirting boards can be removed and refitted, or gaps filled — usually, carpets with a thick underlay will solve this problem, as well as leaks through floor boards.

When insulating, consider environmentally friendly alternatives in your choice of materials. Cellulose fibre is a natural, safe alternative made from recycled newspapers. It is one of the highest-rated insulating materials now available, cutting heating and cooling costs by up to 50 per cent. And not only is the fibre resistant to mice, rats, mildew, rot and fire, but it contains no asbestos or glass and is non-allergenic to human skin.

Insulation should not only be installed in the roof, but where possible in external wall cavities and under floor boards. You are then literally wrapping your house in a thermal blanket. However, wall insulation can only usually be fitted when building a new home or carrying out renovations. Floor insulation does not pose the same problem: the material can be installed beneath floor boards and held in place between bearers by wire netting.

The next area of greatest heat loss and gain are windows. Effective solutions solving these problems are discussed in detail in this chapter.

Your Place in the Sun

Ideally, an elongated house running east to west, with living rooms and kitchen facing north, will maximise living comfortability. Large windows on the north, with wide overhanging eaves or pergolas will give shade from the high summer sun and allow warmth from the lower winter sun.

When buying a house its orientation is not always sited to maximise the benefits of a northern aspect. Preferably, purchase a house that requires minimum renovation to receive plenty of northern light but if this is not possible, and you're prepared to carry out renovations, check the location of the rooms: is it possible to change or add on to expose living rooms to face the north or to install skylights or clerestories?

Building a new home presents a different situation. Plan it with the living, dining, kitchen and family rooms facing north — rooms facing south can still receive winter warmth from the sun by designing your roof to include a clerestory. An attached greenhouse will direct the warmth of the sun into your house in winter, and in summer draw cool air through it.

For maximum solar benefit the orientation of your house should be 10 degrees east of true north. And where possible avoid windows on the east or west aspects, since they can cause your house to overheat in summer; north and south facing windows are best. Even with renovations you can plan your window arrangement; the ideal size being about 15 per cent of the floor area of the rooms, with living rooms up to a maximum of 20 per cent. Too much glass lets winter heat out and summer heat in by conduction, ten times faster than an insulated wall.

MAKING THE MOST OF THE WINTER SUN

Efficient day-time heating can be effectively achieved by getting as much of the sun's warmth inside as possible. Glazing plays an important role in good solar house management. Its thin skin not only admits sunlight, but traps its heat inside a box, room or house. It is by far the simplest solar collector, and the simplest way of using it is by direct solar gains — skylights, clerestories, sunrooms, and attached or window greenhouses.

A solar glasshouse, sunroom or glassed-in porch on a north

wall will quickly warm the inside air of your house, while skylights and clerestories will capture warmth for south facing rooms.

Skylights

A well placed skylight will transform an otherwise dark recess of a room into a bright sunny living area. It will not only give you a view of the sky, but will give life to those ever-delightful indoor plants.

Even a modest 60 cm (2′) square skylight will, on a cloudy day, flood the room below it with a light equivalent to three 100 watt light bulbs. If it faces north it will also contribute to direct-gain heat, and if fitted with a closable vent will allow hot air to escape in summer. However, a northern skylight is best closed off in summer to exclude unwanted heat.

There are a number of commercially manufactured units, designed to suit all needs, that can be installed quite easily by the home handyman. They can be located up to 420 cm (14′) from the floor, without too much loss of light, and must be positioned so as not to be shaded during winter.

The amount of light gain from the skylight is dependent upon its size and orientation, and both of these factors you can certainly control. A good rule of thumb for its size is for its area to be approximately 5 per cent of the floor area to be illuminated. If the room is extremely large it may be better to have two or three smaller skylights rather than one gigantic one — this will also give you far more even light distribution.

For those cold winter nights and hot summer days you will need to construct tight-fitting, insulated louvres to close off the skylight.

The Louvres Make a butt-jointed frame from 50 x 25 mm (2″ x 1″) dressed timber that will fit inside your skylight opening. Divide its length evenly to find out how many louvres you need and what their outside dimension should be. Each louvre should be around 160 mm (6½″) wide.

Next construct the required number of louvre frames from dressed timber 25 x 13 mm (1″ x ½″), making them a little shorter than the width of the skylight frame. Glue and tack 3 mm (⅛″) hardboard on one side, fill the frame with natural cellulose fibre insulation, and fix 3 mm (⅛″) hardboard on remaining side, allowing one of the hardboard sides to extend 10 mm (⅜″) below the frame (this overlap helps to seal the louvres so that warm air won't escape).

Hinge the centre top of each louvre to a 50 x 25 mm (2″ x 1″) slat; fit them into the frame and drill 12 mm (½″) holes through the frame and into the centre of each louvre end and secure with 40 mm (1½″) long dowel pivots. Nail the unit into place.

Close-eyed screws are inserted into the centre of the bottom frame and bottom louvre, and another into the wall just below the frame eye. Thread a cord up through the wall eye and frame eye, then through the louvre eye and back down through the wall eye. Cord ends are tied to a metal cleat screwed to the wall.

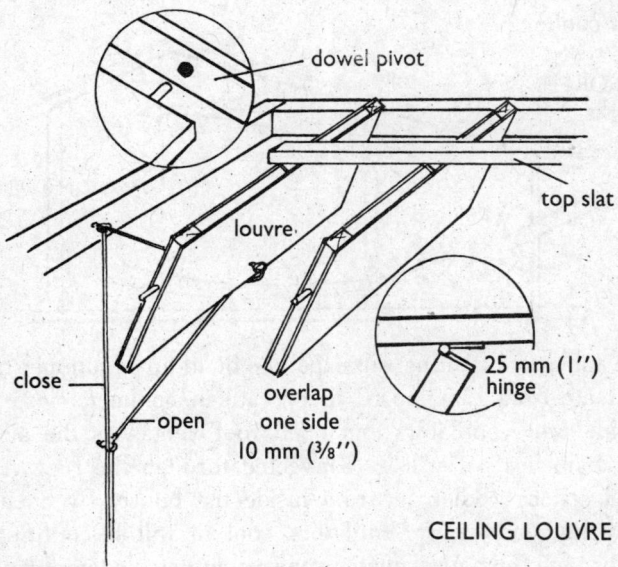

CEILING LOUVRE

Clerestories

Clerestories are really big popped-up skylights. But because of the way in which they are constructed on the roof, they give you a large window with far more solar gain than a skylight. Their vertical, or slightly tilted, glazing will give you more control over when and where sunlight will enter, giving you direct-gain space heating and daylighting that you never imagined possible.

Unless you are building a new home, a clerestory will mean major structural changes to your roof — rafters, ceiling joists and other structural components. Not all existing houses may be able to accommodate such an alteration; cost or complexity may make the project unfeasible, but it is worth careful consideration.

The Sunroom

The sunroom, or glassed-in porch or verandah, is a simple passive solar system which has tremendous potential: it utilises a space that has far too much glass. It should face the north, be attached to rooms that need daytime heating — kitchen, family room and lounge — and be fitted with return and exhaust ventilators. These ventilators should be fitted with shutters to close them off at night, or to be partially closed when the sun room is used for summer cooling.

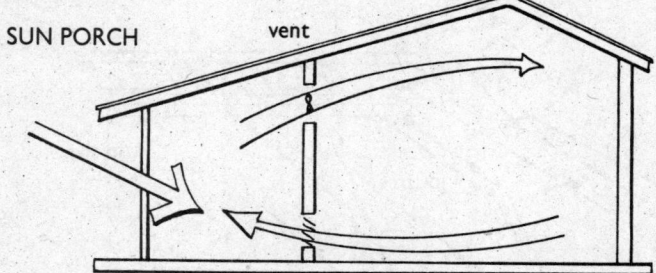

SUN PORCH   vent

Adjustable roof venting will expel the build-up of summer heat and hot air from the house. To operate in summer, close the top inside wall ventilators and open roof vents. As the air in the sunroom heats it will be convected through the roof vents and replaced by cooler air from inside the house. By opening south facing windows, or ventilators, cool air will be continually drawn through the house, maintaining a comfortable temperature.

For winter operation, close the roof vents and open both top and bottom inside wall ventilators. Keep south facing windows or ventilators closed. Again the same principle applies: cooler air is drawn from the house, heated, and, instead of being discharged to the atmosphere, it is redirected into the house. A small, low amperage fan can be fitted in the top ventilators to make them more effective.

Heat Storage Wall

The heat storage wall is a vertical collector with its own storage system. It circulates warm air, by natural convection, into the house during the day and then releases stored heat at night.

This type of system requires a storage wall built of double brick or concrete blocks that have had their hollow cores filled with concrete. The outside of the wall is painted black and framed with glass. An air space of about 15 cm (6″) should be allowed between the masonry and the glazing.

Lower and upper vents are provided for air circulation; a low amperage fan blower, for increased efficiency, can be installed into the upper vent. Fit a thermostat to the fan for automatic shut-off when the collector is cold.

Shutters should be fitted to both top and bottom ventilators to shut off the unit in summer, with the bottom ventilator also being fitted with a one-way flap to prevent the unit from reversing its cycle at night.

Glazing must be vented so that summer heat can escape; small awning type windows, that are lockable from the outside, are ideal. To prevent air from escaping when they are closed, weather-proof stripping is fitted around the inside of the frames.

If the framework fits snugly under the eave, and is protected from moisture and rainwater, flashing will not be required. Otherwise, install a suitable flashing along the top plate.

HEAT STORAGE WALL

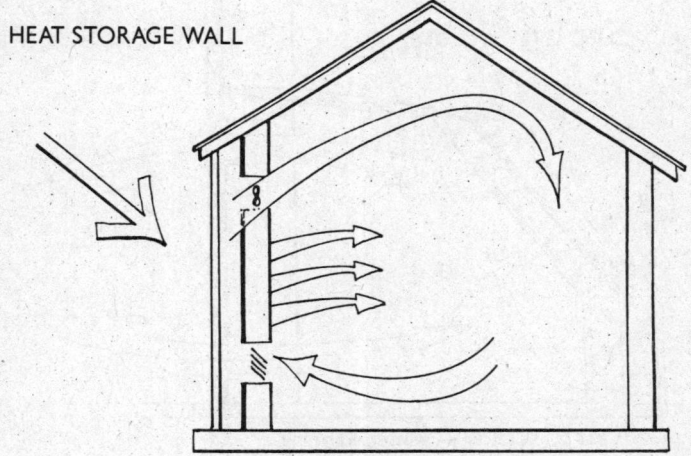

Attached Greenhouse

An attached greenhouse will provide winter warmth, summer cooling, and added living space. And for keen gardeners, space for growing out-of-season vegetables.

Its size depends upon individual needs, but like all solar collectors it should be attached to a north facing wall. Include adjustable vents to trap winter warmth or release summer heat, paint the exterior wall of the house black and install lower and upper ventilators for convective air circulation. Again, inside vents should be shuttered and, if needed, a low amperage fan blower installed in the upper vent.

To increase the effectiveness of your greenhouse or sunroom for summer cooling, add a cool verandah or shade house to the south side of the house. Open adjacent windows or doors during the day and as the air in the greenhouse heats up and is discharged, cool air will be drawn and circulated through the house. At night, simply close off the greenhouse ventilators till the air warms up in tomorrow's sunlight.

A greenhouse need not be an expensive, permanent structure. Prefab kits are readily available where glass panels simply slot into the framework. Additionally, it can also serve as a useful winter entrance hall — acting as an air-lock heat trap, so reducing the likelihood of cold air entering the house with visitors.

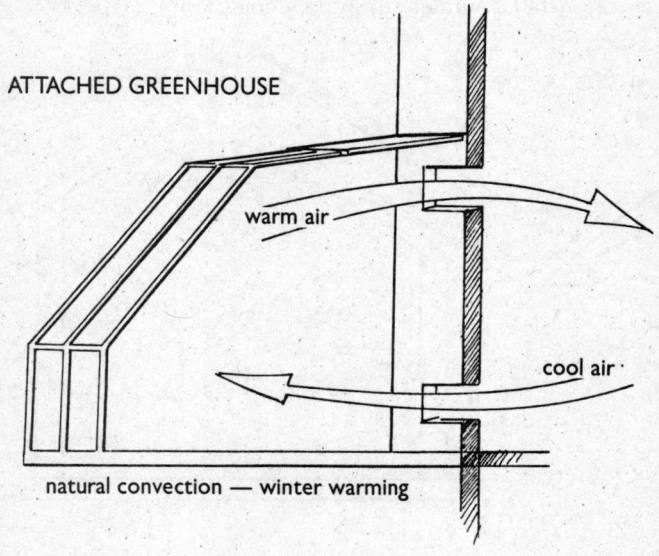

ATTACHED GREENHOUSE

warm air

cool air

natural convection — winter warming

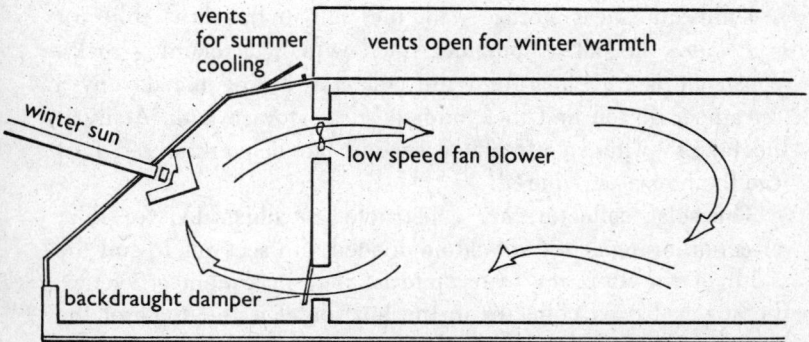

Window Greenhouse

An effective greenhouse can be made in any window opening. It should extend approximately 60 cm (2′) from the windowsill and can be framed and entirely glazed or made from an old window sash.

As well as providing additional winter warmth, it can be used to grow herbs, flowers, and year-round tomatoes. Adjustable shelves will mean extra space for all those different potted plants, and a hanging rod at the top will allow you to dry clothing.

If the unit fits under the eave, flashing may not be required. Otherwise, use a suitable waterproof material such as a piece of galvanised tin plate — available from your local hardware store. Seal any fixing screws or holes with a waterproof silicone sealer.

Sun Traps

Sun traps absorb the warmth of winter sun during the day and then re-radiate the heat at night. Ring your house with large dark pots or sand-filled tubs which will absorb daytime heat, or place them inside the house in front of a north window.

Roof Hot Air Collector

This is a variation of the heat storage wall, and, instead of being a vertical solar collector, uses roof solar absorbers. The storage wall is constructed of stone and releases stored heat at night.

The basic principle of this type of collector is that it stores heat when the sun is at its hottest during the day. Absorber plates need to be quite large, and in areas where frosts are common they should have a surface area of not less than 50 square metres (60 sq yd).

Unlike the heat storage wall, the wall in this heat collector is a hollow insulated container filled with heat-retaining rocks. Warm air is forced around the collector during the day by a low amperage fan and then into the heat storage tank. At night the fan is switched off and a second fan delivers stored heat into the house as required.

The solar collectors are constructed as illustrated for solar air-conditioning. They should be divided into sections to suit the width of the absorber plate material. For maximum efficiency, the angle of each collector to the horizontal is the angle of the latitude plus 10 degrees.

Collector ducts can be either commercially manufactured ducting or can be made from timber and lined with aluminium foil insulation.

The heat storage tank is built of masonry and filled with rocks of varying sizes, but no larger than about 20 cm (8″) diameter. It must stand on a solid foundation, since the total finished weight is going to be several tonnes.

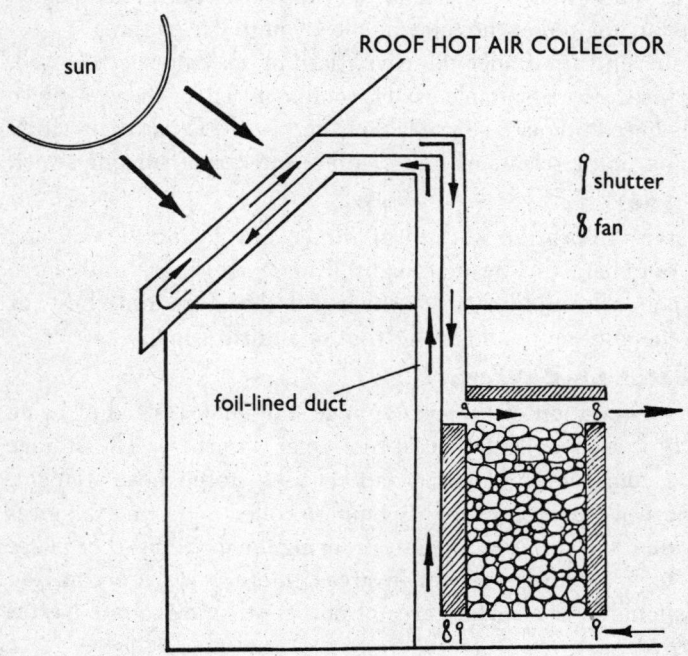

External Modular Wall Air Heater

This type of heater is one of the simplest solar hot air collectors. It can be constructed singly, or in modules, on a north facing wall and will circulate convected air in the same way as a greenhouse. The delivery of heated air can be assisted by a low speed, low amperage fan blower.

The unit is a double-glazed collector box with a corrugated iron absorber panel (aluminium can be substituted for lightness). When the sun's rays strike the absorber plate the air behind it is heated, which then rises and enters the house through a top vent. More air is then replaced through a bottom vent, heated, and the cycle continued.

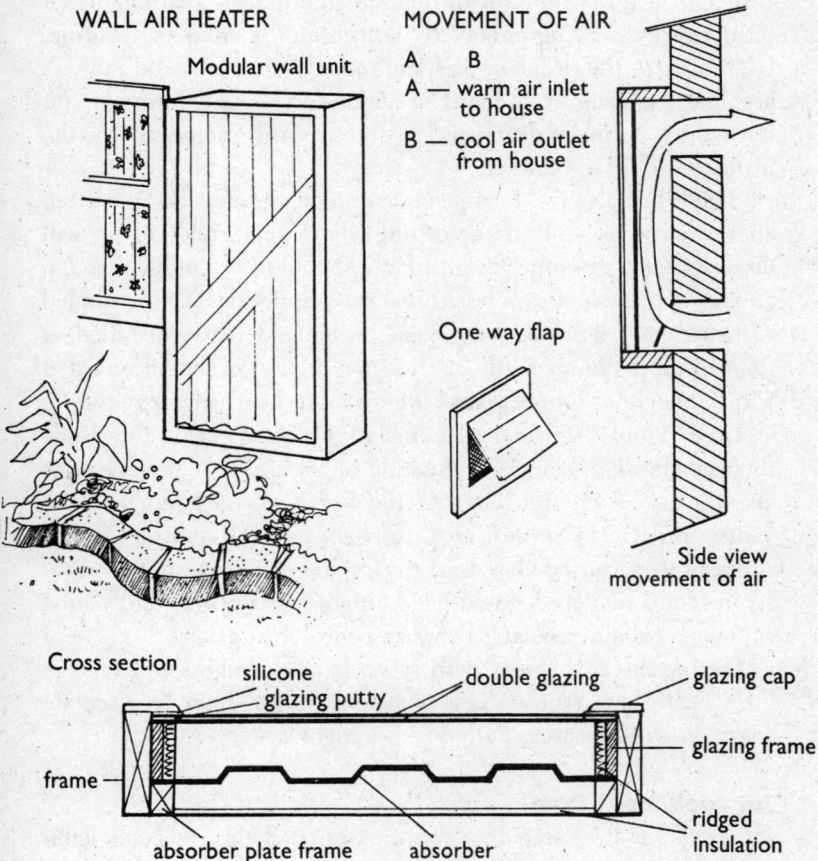

WALL AIR HEATER

Modular wall unit

MOVEMENT OF AIR

A   B

A — warm air inlet to house

B — cool air outlet from house

One way flap

Side view movement of air

Cross section

silicone glazing putty

double glazing

glazing cap

glazing frame

frame

ridged insulation

absorber plate frame

absorber

You can place several of these along the wall for maximum effect, with each one being about the width of the distance between three studs and its length to suit the height of the wall. The absorber plate is painted flat black on both sides, the top of the collector waterproofed with a suitable flashing and storm moulding fitted around its entire perimeter between the frame and wall. Silicone any areas that may present a leakage problem. Fix shutters to both top and bottom vents and a one-way flap to the bottom vent, to prevent the unit from reversing its cycle at night.

DISTRIBUTING WARMTH THROUGH THE HOUSE

Efficient heat distribution throughout the house is an important means of conserving energy by cutting down on space heating. It eliminates the problem of some rooms being hot while others are like ice-boxes. Good air circulation is essential where you have only a single heat source, and is easily achieved by the installation of fan blowers.

Small, inexpensive, low-speed, low amperage electric fans fitted above doorways — near the ceiling where heat builds up — will direct hot air into other rooms. Fans should blow hot air through an appropriate vent in a horizontal direction, and, if not coupled with an open doorway, wall vents will need to be installed at floor level. When deciding upon which doorways should be fitted with fans, choose those rooms where more heat will be required.

In a double storey house install vents between the studs through the floor-ceiling connecting upper and lower rooms. Hot air will rise from the lower to the upper floors through these vents, whilst cold air will drop to be heated. Fit shutters in the vents so they can be closed off during the day, keeping the warm air in rooms that are being used. At night the vents can be opened so that warm air circulates to warm bedrooms upstairs.

During the day, rooms with south facing windows will receive little heat. This type of fan arrangement will help to circulate warm air to these areas both day and night.

Air-lock Heat Trap

In winter avoid using too many external doors. A good habit

to get into would be to use only one. A better alternative is a simple foyer or air-lock: it prevents warm air from escaping, and masses of cold air rushing in, each time you open the door.

It is easily constructed from timber framing, lined both sides with gyp-board, and the cavity insulated with natural cellulose fibre. If the foyer is an external addition, outside cladding should match the outside of the house and the use of insulation is unnecessary. Both doors should be well sealed with weather-proof stripping and a draught excluder.

The air-lock will save heat in two ways: minimal loss of heat around the doors, and a minimal change of air when entering or leaving the house. It is important to remember not to have both doors open at the same time, otherwise your foyer will not do its job.

Install a few shelves and hooks on one wall for coats, shoes, rainwear, or whatever.

SOLAR HOT WATER

Climatic conditions in most parts of Australia make solar hot water heating a sensible, practical and inexpensive approach to providing domestic hot water. It is clean, costs nothing to run and requires no maintenance.

The water heater consists of a fixed flat plate absorber, an insulated storage tank, and insulated connecting pipes. It works on the thermosyphon principle: warm water rises over colder water. As the water in the absorber heats up, its density decreases, and it rises as heavier cold water pushes it up from beneath, and goes along the transfer pipe to the storage tank. This flow continues until the water fails to gain energy from the sun.

The tank must be situated above the level of the absorber for the system to work. Since the hot water cannot seek a higher level, it is stable in the tank until needed.

Commercially manufactured units are readily available in Australia, either with a storage tank mounted directly above the absorber, forming an integral unit, or as two separate components. The latter is far more aesthetically pleasing as it allows the tank to be mounted in the roof out of sight, and can be more effectively insulated.

The Absorber

You can build your own absorber plate unit, however precise and accurate workmanship is absolutely essential for the unit to function. It is far easier to buy a ready-made unit and then adapt it to an existing storage tank. Rheem will supply absorbers only, and though it's not quite do-it-yourself you will save considerable costs on purchasing a complete system.

Storage Tank

A 400 litre (100 gal) tank will give satisfactory storage if the consumption is around 200 litres (50 gal) daily, which is about average usage by a family of four to five. This capacity allows provision for a reasonable carryover for cloudy days.

Insulated copper tanks, such as those commonly used in hot water systems, are ideal and are readily available, new or second hand, if you don't already have a suitable tank. However, an additional inlet and outlet pipe for connection to the absorber plate will be necessary.

The position of the various tank connections is important, and is illustrated in the diagram. The two connections to the absorber must not restrict the flow at the point of entry to the tank. Therefore, to prevent entering cooler water mixing with hot water ready for consumption, the point of entry for the flow connection from the absorber must be sited some distance below the top of the tank. As a rule of thumb, about 90 litres (40 gal) of stored hot water above its point of entry is satisfactory.

Connecting Piping

Since circulation through the absorber is by thermosyphon, and the water must be rapid enough to absorb maximum solar radiation without the temperature rise being excessive and causing high losses, piping between the absorber and the storage tank must be well insulated. Use a high-grade insulation, about 25 mm (1″) thick, that is effectively waterproofed where piping is exposed to the weather.

The size and type of the pipes used is also important: use only copper with a 25 mm (1″) outside diameter, provided the absorber and tank are close to one another. Where long horizontal runs are necessary, the pipe size should be increased.

Installation

Rapid thermosyphon circulation during the daytime requires that both the absorber plate and storage tank be as close together as possible. However, the bottom of the storage tank cannot be closer than 600 mm (2') above the top of the absorber, otherwise reverse circulation will occur at night.

Site the absorber so that it faces due north (though a variation of 10 degrees east or west can be tolerated) and is inclined to the horizontal at an angle equal to the latitude plus 5 degrees to 10 degrees. It must not be shaded by trees or other buildings at any time during the day — summer and winter — between 8 am and 4 pm. An examination of your site at 8 am noon and 4 pm during June and December will reveal any possible obstructions.

Absorber and Tank Arrangement

As already outlined, the best arrangement is where the absorber is sited on the outside of the roof and the storage tank inside. Piping should slope continuously upward from the absorber to the tank, without any humps where pockets of air might collect. Otherwise, thermosyphon flow will be intermittent or may cease to flow altogether.

Auxiliary Heating

There are two ways in which auxiliary heating can be fitted to boost your system during prolonged cloudy weather: a 2 kw electric element fitted to the inside of your tank, together with a thermostat control and time switch. If you are using an existing off-peak electric tank this may already be fitted, and would, therefore, at the most only need minor modification. The other alternative is to hook up the absorber in conjunction with the hot-water jacket of a slow combustion fuel stove or heater. With this arrangement water is automatically drawn to the hottest source, and a continuous supply of hot water will be guaranteed.

Low Pressure System

Both gas and electric hot water systems operate as a high-pressure system, direct from supplied water. However, this type of system is unsuitable for solar water heating, which must be converted to a low-pressure, vented system; otherwise your storage tank may explode.

SOLAR HOT WATER SYSTEM
(coupled with slow combustion stove)

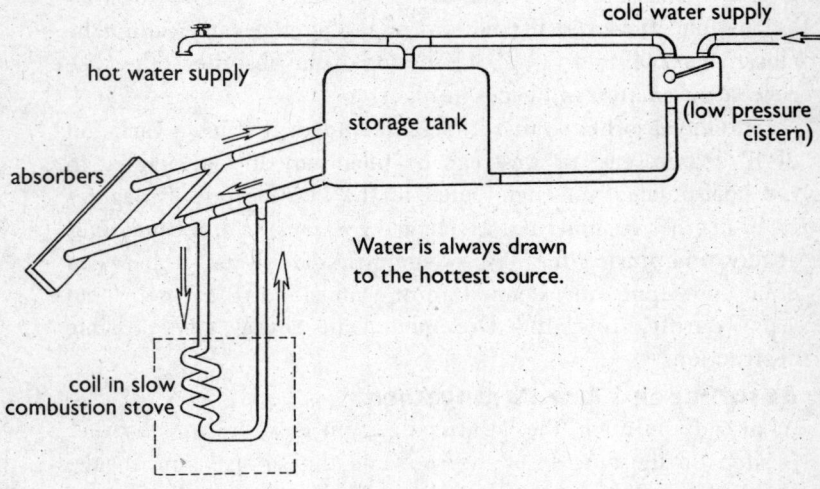

The conversion is simple and only requires the installation of a small reservoir cistern. This operates in a similar manner as the arrangement in a toilet cistern: it has a ball-cock valve which allows water to enter the cistern as water is drawn off to the storage tank. The top of the reservoir tank is left open, and the hot water storage tank, through a T-piece in the hot water delivery pipe, is vented into it.

Frost
People living in cooler areas may have to take steps to protect the absorber against frost. A well insulated cover placed over the absorber at night, and additional insulating around exposed pipes, will help to prevent water from freezing.

Economy
It has been established that two-thirds of the heat energy required in an average year can be supplied by solar radiation. This in fact means that an electric boosted system will reduce your power consumption by two-thirds when compared with an electric off-peak water heater. If your solar heater is coupled with a slow combustion fuel stove or space heater, your power consumption for hot water is nil.

SUMMER COOLING

COOLING SUMMER BREEZES

Ventilation will not only help to keep your house cool, but also increases the sweat evaporation from your skin, giving you a sense of comfort. By directing the flow of summer breezes through the house you will be provided with relief in humid conditions, and the cooling of the house in the evening.

Your house need not face directly into the breeze, as long as it is offset no more than 45 degrees either way; in other cases, a row of shrubs, or high, continuous fence, will redirect the breeze into the house. To gain maximum benefit through this cross-ventilation the breeze should be directed through smaller, low-level openings on the windward side, and then exhausted through larger openings on the downwind side.

SHADING

Because of the summer sun's high angle in the sky it makes the use of pergolas, verandahs, and deciduous plants very attractive for natural cooling; particularly on northern and western aspects.

Temperature control can be as simple as growing deciduous vines, such as grape or wisteria, over a pergola, planting deciduous trees along the northern and western sides of your house, or growing espaliered trees against walls to insulate them in summer and expose them to sunlight in winter. When planting trees that will eventually grow quite large, make sure that they are situated far enough away from the house so that growing branches and roots don't present a problem.

Shadehouses make an attractive and cool retreat from soaring summer temperatures. They can be complex structures of shade cloth and lattice, or again simple in construction like a pergola, with vines growing over it and plants hanging down from it. Light reaching the house is filtered through the greenery and the hanging plants help to lower the surrounding temperature considerably through the moisture evaporating from their containers. This evaporative cooling effect can be further increased by installing a drip system: water lines are attached to the roof framework of your shadehouse above the plants, letting the system drip slowly and constantly. As the day heats up the evaporation increases — more so on extremely hot days — greatly lowering the surrounding air temperature and increasing the cooling effect.

A south facing attached shadehouse, with a water-drip system, combined with a north facing greenhouse or sunroom will give you summer air-conditioning that will keep your home as cool as would a power-hungry mechanical device. The greenhouse acts as a solar heat pump, setting in motion natural convective circulation on the air inside your house. As hot air is vented through its roof, cooler air is drawn from inside the house, and then even cooler air (through strategically placed high vents) from the shadehouse.

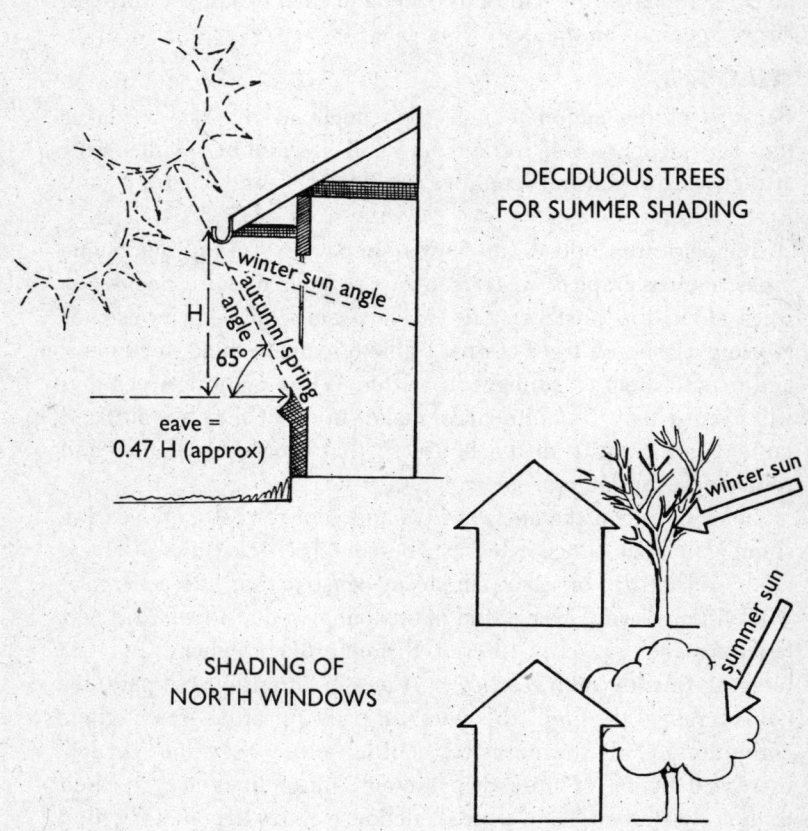

DECIDUOUS TREES
FOR SUMMER SHADING

winter sun angle

autumn/spring angle

H

65°

eave =
0.47 H (approx)

SHADING OF
NORTH WINDOWS

winter sun

summer sun

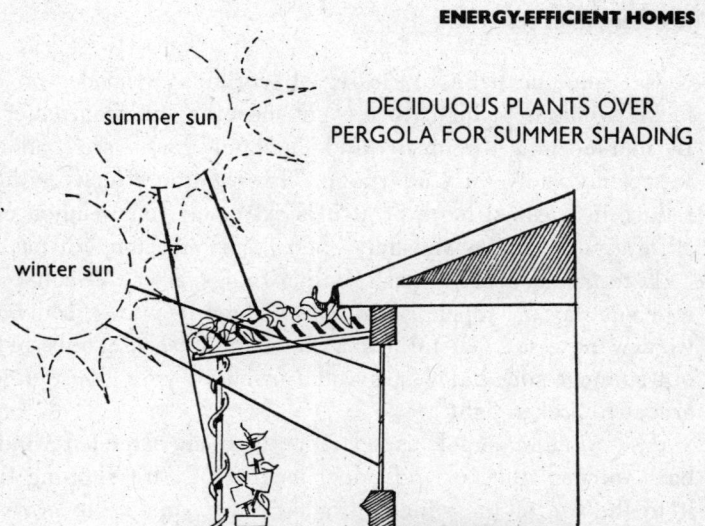

summer sun

DECIDUOUS PLANTS OVER
PERGOLA FOR SUMMER SHADING

winter sun

Window Shading

Solar heat gain through windows and sliding glass doors can be greatly reduced if drapes, blinds, shutters and shades are closed when the sun is shining on them. To ensure maximum protection, internal shades must be well sealed on the bottom and sides, and on the top by a pelmet. Without the pelmet, curtains will increase heat gain by settng up a convective loop.

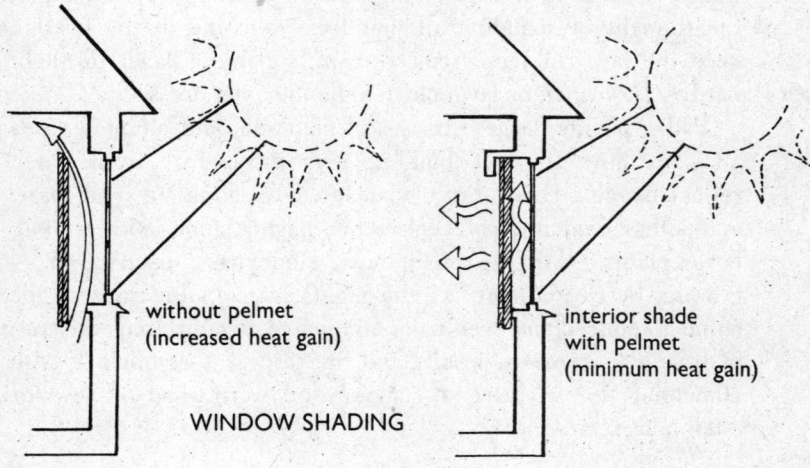

without pelmet
(increased heat gain)

interior shade
with pelmet
(minimum heat gain)

WINDOW SHADING

Get into the habit of closing all curtains or shades on east facing windows at night to prevent morning sun from entering. By mid-morning the north windows should be covered, and the west windows by early afternoon. This rotation effectively helps reduce heat gains; however, if it is extremely hot weather, close all drapes or shades and only open them once the sun has set.

External shades, such as roll-up awnings, are an effective and efficient way of stopping heat gain through windows. Individual window pergolas will form a living blind that not only blocks out summer sun, but will give the inside of your house a cool, greenish, broken light.

The window pergola is no more than an extended window-box, with a lattice or individual lengths of wire running from it to the top of the window frame. Plants suitable to grow are evergreens like passion fruit, for hot areas where you wish to block the sun all year round, or deciduous vines such as grapes which allow the entry of winter sun. Summer crops like climbing beans also work well, providing the additional benefit of a vegetable and then leaving the window-box free for a splash of colour from winter flowers.

Non-reflectors

The good old Aussie suburban lawn has a lot going for it as a climate modifier. Combined with other greenery around your house it has a definite cooling effect. And if you shudder at the thought of spending all that time sweating in the heat to keep the lawn in trim, replace it with a thyme lawn (drought hardy), *Dichonrad*, or lawn chamomile in cooler areas.

Paths, paving, large expanses of concrete, and bitumen roads all contribute to heat build-up and discomfort from harsh reflected light. Hedges can be planted to eliminate road glare, and paths eliminated and replaced by paving stones with ground-cover plants — thyme, chamomile, moneywort, pennyroyal — growing between them. As the plants spread, the glare of the stone is reduced and eventually eliminated. Try different varieties of thyme — caraway, woolly and Shakespear — combined with chamomile for a variety of colour, and wormwood or yew for a tall hedge.

Pots of Water

This is, without doubt, one of the easiest methods of cooling, relying on evaporation for instant water cooled air-conditioning. Select a few medium to large round goldfish bowls, add a layer of dirt, root in some water plants, fill with water, and place them below windows where the draft will flow over them.

Fish ponds located beneath pergolas and in a line close to access doors will achieve the same effect, as will large ground pots that can hold water.

REDUCING INTERNAL HEAT GAIN

Heat generated by people, lights and appliances all greatly contribute to your cooling systems being ineffective. By modifying your household habits this heat gain can be reduced.

- try to prepare meals that use minimum stove cooking time, or require no cooking at all
- if you do intend to cook, use a microwave oven as much as possible or adapt those energy saving ideas in chapter 1, Cooking
- avoid using the oven — if baking, do this in the cool of the evening when breezes can blow away heat, or the early morning before the heat of the day starts
- refrigerators and freezers both give off heat. If possible, put them in a garage, laundry, or on a verandah

Simple solar modifications will help: install a screened, closeable vent in the floor behind the appliance and a vent in the ceiling above it connected to a length of copper or stainless steel flue pipe. The flue should extend about a metre (yard) above the roof where it exits, and be fitted with a screened weatherproof cap.

Hot air from the appliance is exhausted up the flue by convection, and cooler air to replace it is drawn from under the house. During the heat of the day the flue will heat up and function as a small heat pump, making the system even more effective.

Refrigeration appliances will operate more efficiently if kept a little away from the wall. The coils should be kept clean and the doors should be opened as little as possible.

ROOF COOLING AND VENTING

Ceiling insulation significantly reduces the heat gain from a hot roof. However, this can be further reduced by the following roof treatments:

- avoid dark roofing materials as they only absorb heat; use a light colour instead
- iron roofs, such as found in most rural areas, greatly reflect the sun's rays
- roof and eave venting will reduce the inner heat of the roof space. A continuous ridge vent with soffit air intake is by far the best venting system. If this is not practical, a vented cupola or suitable single roof vent will help. But to make them really effective you will still need soffit vents

ROOF AND EAVE VENTING

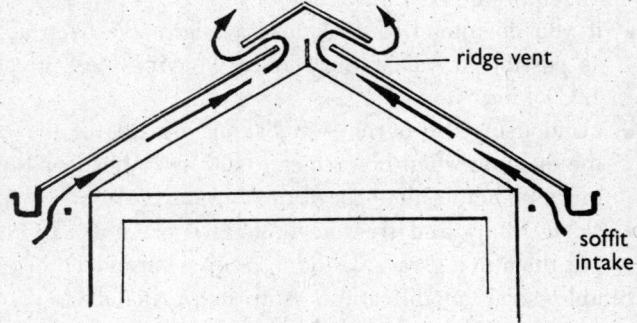

ridge vent

soffit intake

SOLAR AIR-CONDITIONING

Solar air-conditioning is a simple and effective means of natural cooling. Like the attached greenhouse, it is a passive system that works entirely by natural convective ventilation that is boosted by the power of the sun. It is easily constructed and installed, and is an alternative when it is impractical or undesirable to construct a more elaborate structure.

The solar unit, which is similar in appearance to a hot water absorber plate, sits on the roof parallel to and just below the line of the ridge. Its inlet opening (OB of the diagram) is connected by ducting to a **closeable** vent in the ceiling. Cool air is drawn through the house from outside via vents in a south

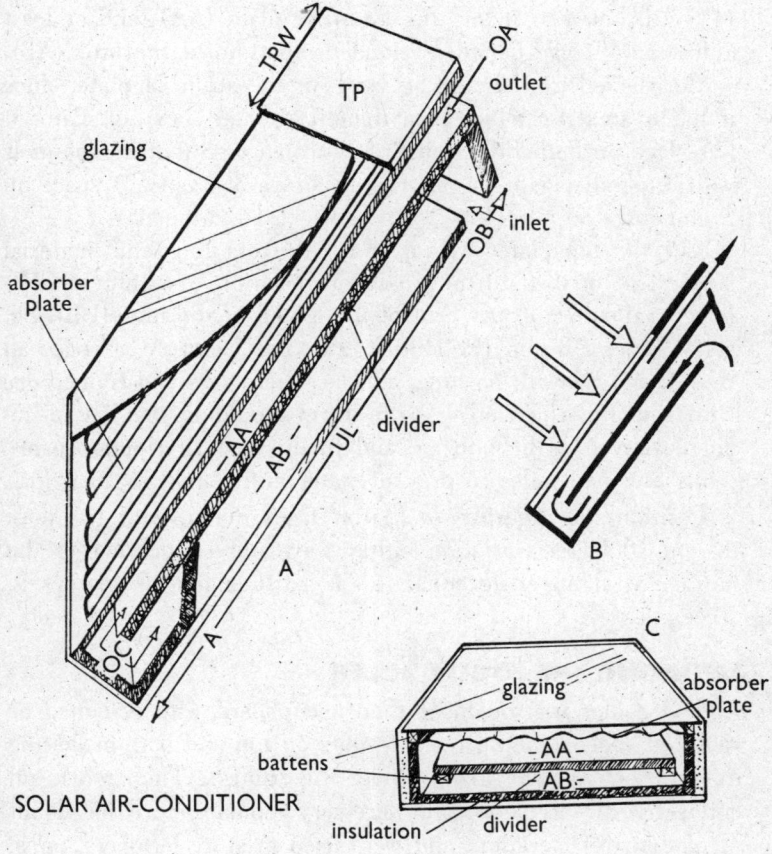

SOLAR AIR-CONDITIONER

facing wall, adjacent to a shadehouse, or preferably from vents in the floor, at the rear of the house, connected to the crawl space.

Construct the roof unit from 15 mm (³⁄₅″) thick, exterior grade, waterproof plywood; you will require one base and four sides, glued and nailed for added strength. Affix battens along the inner frame edge to support both the channel divider and the absorber plate (diagram C). Line the inside base and either side of the channel divider with rigid foam insulation — this is important in the reduction of heat loss while the system is operating. Fix the channel divider, which is 15 cm (6″) shorter than the length of the unit (diagram A) and has at least a 10 cm

(4″) gap between it and the absorber plate (AA), and at least a 10 cm (4″) gap between it and the bottom of the unit (AB).

Corrugated iron should be used for the absorber plate, since it has a greater surface area than flat metal sheeting. Paint it flat black on both sides and fit it into place so that its channels will be parallel with the air flow. Allow a 3 mm (⅛″) space all round and a 30 mm (1³⁄₁₆″) gap between it and the glass.

Cut the top plate (TP, diagram A) from the same material as the rest of the unit and nail into position; its width (TPW) is approximately 10 per cent of the length of the unit (UL). Set glass, at least 3 mm (⅛″) thick, allowing 2 mm (⅛″) space all round, and fix with beading. Seal between glass and frame edge with silicone sealer, cover with storm moulding that has a 45-degree bevel to the outside, and finally seal all cracks, seams, joints and nail holes to prevent water entry and any heat loss.

Position the absorber so that it faces north, with no more than a 10-degree variation either way, and is inclined to the horizontal at an angle equal to the latitude plus 5 degrees to 10 degrees.

UPDATING THE FOOD COOLER

A food cooler was no more than a cupboard with screened or slattered shelves, and screened vents on top and bottom, letting in cool air, usually from under the house. They work on convective air circulation and were very popular before the advent of modern refrigeration, and were used to store leftovers, jams, fruit, onions or potatoes.

When I moved to live in the country the house I purchased had one of these old-fashioned coolers attached to the outside kitchen wall. However, what it needed was a great deal of imagination to turn it into something more than just a vegetable storage cupboard. What I finally came up with was a combination of it and the well known Coolgardie Safe.

The inside of the cupboard, behind the slattered cooling vents, was lined with a stainless steel insect-proof mesh, and the inside base, roof and door were lined with a rigid insulation and then Laminex to prevent any unnecessary heat gain. A water cistern, designed to work on the same principle as an automatic chicken waterer, and fitted with a ball-cock valve to regulate water flow, was installed on top of it. Around its base perimeter a narrow gutter was connected, with three lots of heavy duty canvas, each one consisting of three layers of canvas sewn together, tacked to each side. One end of the canvas was placed in the cistern water trough and the other end in the lower collection gutter.

Although it may not be everyone's idea of a necessary house addition, it is a marvellously simple and effective system for keeping perishable items, such as butter, milk and cheese, cool and fresh. Fruit and vegetables store equally well.

The principle on which the unit works is the same as any refrigeration plant: evaporative cooling. Water continually flows through the canvas, evaporating as it goes to provide a cool atmosphere around the stored food. The lower collection gutter diverts water through a drip-pipe outlet to anywhere you want it: storage tank, garden plants, etc.

Although an addition of this type would not replace a domestic refrigerator, it would allow the use of a much smaller unit and mean a greater saving in energy. And for city and suburban dwellers, the Coolgardie cupboard could quite easily be designed to blend in with the aesthetics of your house.

SOLAR ELECTRICITY IN THE HOME

For those people building a new home solar electricity may be the answer. It is possible to have a system that will provide all your power needs, with the exception of an electric stove, and be comparable in installation costs to that charged for connection to the main power grid. Check out all costs thoroughly before proceeding, and especially the cash rebates now available for going solar — they can be quite substantial.

Even for existing homes solar electricity may provide an energy-efficient juxtaposition with existing systems. A small solar array and storage batteries could run internal lighting, outside lighting — garden, swimming pool, barbecue, porch or verandah — swimming pool filter pumps, internal ceiling fans, exhaust fans and the like, all of which would mean a little less dependence on fossil fuels.

CLEANING & CLEANSING   3

Cooking oils, used or unused, and fat can easily be turned into safe, biodegradable soap to suit many different household purposes. Making your own soap products will not only help our environment, but will save you money as well and is good, clean fun.

Soap is the result of a caustic soda solution or lye (alkali) acting upon fat or oil. When washing, it is the fat that makes a lather of suds to loosen dirt, while the lye breaks down the resistance of stubborn grease; water then dissolves the lye and washes away the dirt.

Homemade soap contains no harmful phosphates or detergents, both of which are harmful to the environment, in particular encouraging the growth of algae and consequent pollution in drainage systems and waterways. Unlike many commercially manufactured soap, suds from the homemade variety are safe to use for watering garden plants and will help to deter many insect pests, especially aphids.

RECIPES

To simplify the recipes, the following abbreviations have been used in the text:

g	grams	*tbsp*	tablespoon
ml	millilitres	*dsp*	dessertspoon
l	litres	*tsp*	teaspoon

INGREDIENTS IN HOMEMADE SOAP

Lye

A good lye of average strength can be made by dissolving caustic soda in the proportions of 900 g (2 lb) to each 4½ litres (1 gal) of water, or as directed for a particular recipe. However, you can make your own lye by the following method.

Drill a few holes in the bottom of a plastic garbage bin and then stand it on blocks in a large enamel or ceramic collection dish. Lay some straw in the bottom of the bin and fill it with finely sifted wood-ash, but not the kind from paper wastes as they often contain clay and other additives, and pour a bucket of cold rainwater or distilled water on top of the ash. Repeat this process every three hours on the first, third and fifth day. The water that drips out the bottom will be lye.

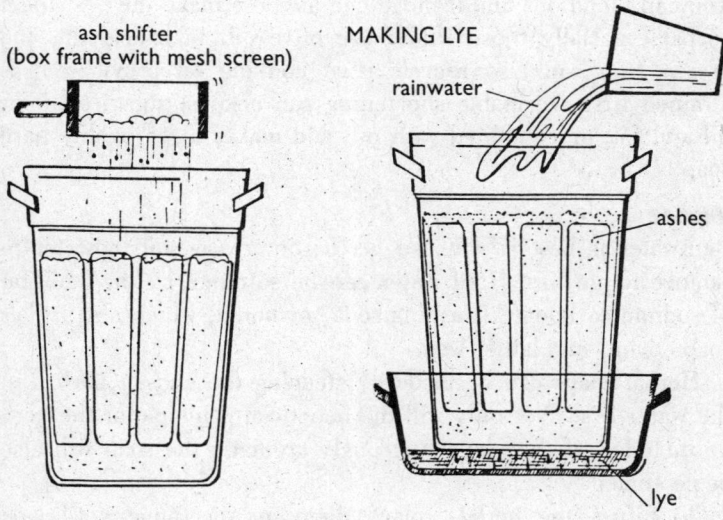

ash shifter
(box frame with mesh screen)

MAKING LYE

rainwater

ashes

lye

WARNING: Lye is highly caustic and should be handled with caution at all times. Don't allow children to go anywhere near it, and as an added precaution, be sure to wear rubber gloves when handling it. When adding the lye to the soap mixture pour it in slowly to avoid spattering and contact with the skin and eyes. Keep your face well away from the pouring lye, as an alkaline burn to the eyes means permanent loss of sight. If a burn to the skin should occur, immediately wash the affected area with cold water and keep covered with a wet, cold cloth while you seek medical help. Should any lye come in contact with the eyes rinse immediately with clean water and seek medical help promptly.

Fat

Clarified beef fat (tallow) makes the hardest soap, although bacon and sausage grease and dripping from cooking works equally well. You can buy clean, white rendered lard from most supermarkets and it is inexpensive and makes good soap.

Vegetable oils, such as olive, sunflower, soya and safflower, can be used alone or in combination with each other or blended with coconut oil, palm-nut oil, linseed and castor oil. However, most oils are expensive to buy, with the exception of sunflower oil, and generally make a soft soap which doesn't last very long. You can blend oils and fat together, as these make the best toilet soaps; especially those which use olive oil. Like dripping and grease, you can also recycle used cooking oil provided it is strained first. Vegetable shortening can also be substituted for oil and fat, or combined with oil, and makes a reasonably hard soap.

Water

Rainwater is best if you can get it, otherwise boil tapwater to remove impurities. Hard water can be softened by the addition of sodium carbonate (washing soda) or borax, which also makes soap creamy and lather well.

Herbal soaps can be made by steeping the chosen herb/s in the water first. Not only will the beneficial principle of the herb be included, if the plant is strongly aromatic the soap will also be perfumed.

To infuse the herb/s, place them in a ceramic bowl, add required amount of boiling water, cover, steep overnight, and strain through muslin cloth. The following proportions apply:

8 tbsp (10½ US tbsp) fresh leaves or flowers, or

2 tsp dried leaves or flowers

to every 600 ml (20 fl oz) of boiling water.

Although the choice of herbs is entirely up to you, the following will give your soap an inviting, natural fragrance:

lavender	aromatic fresh scent of lavender
pine needles	fresh pine scent
lemon verbena	delicious lemon scent

rose petals	delicate scent of rose
rosemary	refreshing, soft aromatic fragrance
rue	strong aromatic fragrance

Other herbs can also be included for the beneficial effect that they exert upon the skin.

chamomile	slightly antiseptic; particularly beneficial to people with oily skin, especially when used with equal parts of rosemary
lavender	has antiseptic and disinfecting properties
lovage	natural deodorant qualities
malva	combine with marshmallow and thyme for acne
plaintain	healing; use for skin ailments
rose petals	softens the skin
rosemary	softens the skin and is a natural deodorant
thyme	helps to clear spots and acne
yarrow	good for oily and greasy skin

Scent

To scent the soap, add a few drops each of the appropriate essential flower or herb oils just before pouring into the moulds, making sure that the oil is well mixed. Try any of the following: cloves, cinnamon, eucalyptus (use sparingly as it has a powerful fragrance), geranium (all types), jasmine, lavender, lemon (all types), lemongrass, lime, melissa (lemon balm), neroli, patchouli, peppermint, pine, rose, rosemary, sage (all types), spearmint, tangerine, thyme and verbena.

Colour

I prefer to have a clean, white colour. If you must colour it, do so by adding a natural dye after the lye has been added. Be careful though, otherwise you may end up with a different shade to your normal skin colour. Saffron, turmeric and caramel will give you various shades of yellow.

EQUIPMENT FOR SUCCESSFUL SOAP MAKING

You will need:

- a large mixing bowl made of glass or ceramic, and a large enamel cooking pot. Under no circumstances use bowls, pots and utensils made of aluminium, iron, tin or plastic — the caustic lye will eat holes in them
- a wooden spoon for mixing and stirring the mixture
- moulds for the soap mixture. These should be shallow containers, such as enamel or stainless steel pans or wooden or cardboard boxes, lined with damp calico or muslin; a depth of about 5-7 cm (2"-2¾") is ideal. For fancy cakes of soap use a variety of sizes and shapes for moulds — patty pans; chocolate moulds; rubber plaster of paris moulds; circles, triangles, etc, using cardboard. Again line with wet muslin or waxed paper
- rubber gloves for handling caustic lye solution
- length of piano wire attached to two pieces of dowel, or a large knife, to cut the soap into bars

Storing the Soap

Soap should be left to harden and cure for a few weeks to ensure that any free caustic soda is incorporated. In fact, the longer you leave it the harder it becomes and the longer it will last.

Borax

For a creamy soap, borax can be added to the basic soap recipes in the proportion of 1-10 by volume. Where a basic recipe is being used as a base for another recipe, exclude the borax unless otherwise directed.

Soapmaking Hints

- always add caustic soda to water
- add caustic lye solution carefully to the fat or oil in a thin, slow stream
- wear rubber gloves when handling caustic soda or lye solution
- pour the mixture into moulds once it is cool and has a honey-like consistency

- store soap in a stack that allows air to circulate whilst curing

Clarifying Fat

Heat fat in an enamel saucepan using just sufficient heat to melt it. Skim off any large pieces of meat and then strain through muslin cloth. Old-fashioned type fat strainers can still be bought from most major supermarkets and kitchen specialty shops.

Removing Salt from Fat

Place meat or bacon fat in a saucpan and cover with water. Bring to the boil, remove from heat, and allow to cool. When cool fat can be skimmed off and meat or crackling scraped from the bottom. Return to a clean pan and clarify and strain.

By boiling the fat in water, salt and other impurities sink to the bottom of the saucepan.

Softening Water

Hard water makes lousy soap! To soften it, add washing soda or 1 tablespoon (1⅓ US tbsp) of caustic soda to 4 litres (1 gal) of rainwater or boiled tapwater, stirring constantly. Allow the mixture to stand for about one week, and then pour off the water from the top; this makes excellent soft water for soapmaking.

Basic Hard Soap

>   *500 g (17¾ oz) caustic soda*
>
>   *2 l (4¼ pt) soft water*
>
>   *3 kg (6⅔ lb) salt free, clarified fat*

Put the water in a large ceramic bowl or enamel pot and add the caustic soda, stirring until completely dissolved. The mixture becomes extremely hot and must be set aside until lukewarm.

At the same time melt the fat and allow it to cool but not solidify again. Add the caustic solution to the fat and stir for 5 minutes. Pour into moulds and keep in a warm spot for 24 hours, cut into bars, and then cure for several weeks.

Basic Soft Soap

>   *500 g (17¾ oz) caustic soda*
>   *1½ l (3 pt) soft water*
>   *1½ l (3 pt) vegetable or seed oil*

Prepare as for basic hard soap, adding caustic solution to unheated oil.

Reasons for Failure

Separation	■ too cold or too hot a temperature used
	■ soap is too vigorously or not thoroughly mixed
	■ exceeding rancid fat, or fat containing salt used
Greasy layer on top	■ too much fat or oil for the amount of caustic
White deposit on the soap	■ hard water used in making the caustic solution
	■ a little free caustic
	■ the addition of too much borax
Streaking soap	■ a thorough mix of ingredients wasn't achieved
Cracks in the soap	■ too much stirring
	■ too much free caustic
	■ drying too quickly

Correcting Soapmaking Problems

- if the soap does not mix properly and remains greasy and streaky, put it in a dish of cold water and continue stirring until thick. Pour into moulds
- if the temperature of the ingredients is too low and they will not mix and go lumpy, place in a dish of warm-to-hot water and gently stir until it is the right consistency and all the caustic is incorporated
- if, after the soap has been moulded, it separates, break it up (wearing rubber gloves), place it in an enamel pot, and reboil each ½ kilo (17¾ oz) to 400 ml (13 fl oz) of water. Boil until the mixture drops in sheets; if it won't work, add more water and stir until it does

IN THE BATHROOM

Toilet Soap

A good toilet soap makes the outer layer of the skin, the so-called horny (epithelial) layer, smooth and soft.

Soft White Soap

    250 g (9 oz) caustic soda   *1½ l (3 pt) coconut oil*

    1 l (2 pt) soft water   *50 ml (1²/₃ fl oz) castor oil*

Prepare the same way as basic soap recipe.

Coconut oil may be replaced by pure olive oil, and the water portion added as a herbal infusion. When adding a herbal infusion, prepare the caustic solution using 60 per cent of the soft water and then add the remainder as an infusion when mixing all ingredients together.

Creamy White Soap

This soap recipe is based upon the famous Castile soap, using a combination of oil and fat. It produces a hard, creamy soap that feels soft and smooth against the skin.

The olive oil in the recipe can be replaced with any other vegetable oil, however it will not be quite as creamy; although its cleansing properties will be in no way impaired.

    350 g (12½ oz) caustic soda   *750 ml (25 fl oz) olive oil*

    1 l (2 pt) soft water   *2 kg (4½ lbs) tallow*

Prepare as you would for the basic soap recipe. Cut into bars as soon as the mixture sets in the mould, otherwise it will be brittle.

Shaving Soap

This recipe makes a subtly fragrant soap, ideal for shaving.

    1 cake creamy white soap,   *2 drops essential oil of thyme*

    grated   *1 drop essential oil of*

    distilled water   *peppermint*

    4 drops essential oil of   *5 drops essential oil of*

    lavender   *bergamot*

Melt soap in an enamel pan, over a low heat with just enough water to form a soft paste (use a potato masher to help dissolve soap). Stir in essential oils until well blended, remove from heat and spoon into a wide-mouthed, shallow container. Allow soap to harden for 48 hours before use.

All the essential oils may be omitted from the recipe and the fragrance included as an infusion of the herbs with the distilled water (see Flower and Herb Waters, chapter 8).

Shaving Cream
This soft cream is excellent for those who prefer not to use a shaving brush and lather-up.

> 1/2 *cake creamy white soap,*
> *grated*
> *distilled water*
>
> *175 ml (5³/₄ fl oz) rosewater*
> *175 ml (5³/₄ fl oz) vodka*

Melt soap in an enamel pan, over a low heat with just enough water to form a soft paste when cold. Dissolve rosewater in alcohol and mix with soap paste. Store in a wide-mouthed, screwtop jar.

Apply to face with fingers, and smooth over beard.

Transparent Soap
A special soap for that really luxurious feeling, and for those people who prefer not to use animal products.

All the ingredients in this recipe are measured in parts by weight (pbw), and are shown:

> *35 pbw coconut oil*
> *25 pbw sunflower oil*
> *11¹/₂ pbw castor oil*
> *35¹/₂ pbw caustic soda*
> *solution*
>
> *20 pbw sugar*
> *15 pbw soft water*
> *20 pbw alcohol*
> *herbal oils*

Mix together vegetable oils in a double saucepan, warming to 44°C/111°F. Stir in warmed caustic soda solution, ensuring that it is at the same temperature, cover and simmer for one hour, allowing mixture to steam gently. Dissolve sugar in water and set aside. After an hour a clear, grain-like soap will be seen in the pan; add the sugar solution and alcohol and mix well. Replace the lid and cook for a further 10 to 15 minutes or until the mixture has sufficiently saponified, i.e. when it becomes thick and light in colour. Add a few drops of fragrant oil, according to preference, mix well and pour into moulds. Allow to cure before using.

To make a caustic soda solution, dissolve 450 g (1 lb) of caustic soda to each 2¼ litres (4²/₃ pts) of soft water.

The alcohol content of the soap can be vodka, or a denatured alcohol such as methylated spirits. To mask the smell of the methylated spirits, dissolve the sugar in an infusion of a strongly aromatic herb and adjust the final scent with a fragrant oil of the same plant.

Honey Soap
This is particularly good for people with dry or sensitive skins. You can use either one of the basic soap recipes or the creamy white soap as its base.

900 g (2 lb) grated, unscented soap ·

113 g (4 oz) unprocessed honey (from your health food store)

113 g (4 oz) almond oil

113 g (4 oz) purified borax

essential oil of your choice for fragrance

Place grated soap in the top of a double boiler with barely enough hot water to cover. Melt over boiling water, stir thoroughly and add the honey, almond oil and powdered borax. Mix well by continuing to stir for about ten minutes, then add a few drops of fragrant oil for scent and remove from heat. Pour into moulds, allow to harden, and cut into cakes. This soap can be used immediately as it has already been cured.

Cosmetic Herb Soap
An ideal soap to use on oily or problem skin, which again uses either the basic soaps or the creamy white soap as a base.

450 g (1 lb) grated, unscented soap

250 ml (8 fl oz) elder flower and yarrow infusion (see Flower and Herb Waters, chapter 8)

225 g (8 oz) almond meal

purified borax

Put the grated soap and herb infusion in a double boiler and dissolve together over a medium heat. To this mixture add borax — in the proportion of one part to ten by volume — and almond meal. Mix well by stirring for about ten minutes. Remove from heat and pour into moulds; allow to harden, cut into cakes, and use immediately.

Oatmeal Soap
Add 2 cups of medium oatmeal to the basic soap, or creamy

white soap, mixture just before it is poured into the moulds, making sure that it is well mixed.

Floral Washballs

300 g (10¾ oz) basic soap

300 ml (10 fl oz) rose water

10 drops oil of cloves

3 tbsp (4 US tbsp) dried lavender flowers

3 tbsp (4 US tbsp) dried rose petals

2 tbsp (2⅔ US tbsp) dried sweet marjoram

Grate the basic soap into a large ceramic mixing bowl. Heat 250 ml (8 fl oz) of rose water to just below boiling point and pour it over the soap. Blend thoroughly with a wooden spoon and allow to stand for 10 minutes. Knead with your hands to make a smooth paste, and mix in the oil of cloves, lavender, rose petals and marjoram. Leave in a warm spot for 10 minutes or until the mixture begins to dry and becomes pliable. Form the soap into small balls, about the size of a golf ball, and leave in the sun, on a sheet of plastic film, for about 2 hours to firm up.

Moisten your hands with the remaining rose water and rub the balls to make them smooth and shiny, then place them back on the plastic film and leave in a warm spot for 24 hours to firm completely.

Soapwort Bathing Liquid

40 g (1½ oz) dried soapwort root or leaves

600 ml (20 fl oz) water

Put the herb and water in an enamel saucepan and bring to the boil; allow to boil for 4 minutes. Remove from heat, cover and steep for 30 minutes. Strain through muslin cloth, squeezing any remaining liquid from the herb, and store in a tightly sealed bottle in a cool place.

Half a cup of liquid added to bath water will give it a gentle, cleansing effect that leaves the skin feeling smooth and soft.

Soapwort Shower Gel

40 g (1½ oz) dried soapwort root or leaves

600 ml (20 fl oz) water

15 ml (½ fl oz) vodka

arrowroot

Prepare the soapwort liquid in the same way as the washing liquid. After straining, measure the liquid, return to the enamel saucepan, add the vodka, and 2 teaspoons of powdered arrowroot to every 80 ml (2⅔ fl oz) of liquid. Heat gently, stirring continuously until the mixture thickens and clears. Remove from heat and store in a wide-mouthed container.

Rub the gel over your body whilst showering for a gentle cleansing effect that leaves the skin soft and smooth.

Soap Bubble Liquid
This is a great way for kids to enjoy their bath, at the same time washing away dirt and grime.

2 parts grated basic hard soap	*40 parts soft water*
	aromatic dried herb of your
30 parts glycerine	*choice*

Put the dried herb — in the proportion of 2 tsp to every 300 ml (10 fl oz) of boiling water — in a ceramic bowl and add the soft water. Cover, steep overnight, and strain through muslin cloth.

Add the grated soap and herb water to an enamel saucepan and dissolve together over a medium heat, stirring continually. Stir in the glycerine until well blended, remove from heat, cool, and store in a tightly capped bottle.

Pour a small amount into the bath while the taps are running, and swish around to create the bubbles.

SOAP FOR THE KITCHEN

Herbal Dishwashing Liquid
This is a biodegradable liquid soap that will remove grease and grime from your dishes and still be gentle on your hands.

60 pbw olive oil	*alcohol and water, sufficient of each*
15 pbw potassium carbonate (potash)	

Dissolve each 15 parts of potash in 30 ml (1 fl oz) of water, heat the oil in an enamel double boiler, add the potash solution (previously warmed), and stir briskly. Continue a medium heat until the mixture has thickened to the consistency of thick pea soup. If oil globules separate out and refuse to thicken, the potash is not the correct strength, and more must be added — 1 or 2 parts dissolved in water. Remove from heat and allow to cool.

The soap made will be like jelly; it is dissolved in alcohol, 135-175 ml (4½-5¾ fl oz) of soap to 55 ml (1¾ fl oz) of alcohol. Stand for two days then drip through filter paper, and then add fragrant oil, such as lemon or lime, if desired.

You may add soapwort, by first preparing a decoction (see chapter 8), and then dissolving the potash in the herb water.

Alcohol can be either vodka or methylated spirits. If you don't like the smell of the spirits (denatured alcohol), this can be disguised by the fragrant oil and the addition of the same herb in the decoction.

Herbal Dishwashing Gel

½ cake basic soap, grated

4 tbsp (5⅓ US tbsp) dried soapwort roots or leaves

4 tbsp (5⅓ US tbsp) dried lemon verbena

9 litres (2¼ gal) of water

½ cup washing soda

Place the dried herbs in a large plastic bucket and add 8 litres (2 gal) of hot water. Cover and allow to steep overnight; then strain through muslin cloth, squeezing all the liquid from the herbs.

Add the grated soap, and a little of the reserved water, to a saucepan. Bring to the boil, stirring constantly until the soap has dissolved (use a potato masher if necessary). Boil the remainder of the water, add this, plus the washing soda, to another bucket, stirring until dissolved; then add the soap solution and herbal infusion. The mixture will set into a soft gel; use ¼ to ½ cup for washing dishes. If you include more grated soap in this recipe, the quantity needed will be less — experiment to determine your own requirements.

Store in a lidded bucket.

LAUNDRY NEEDS

Pure Laundry Soap

This recipe will give you an all-purpose good, hard laundry soap.

20 pbw clean lard or tallow

soft water

> 10 parts pbw caustic soda   *common salt solution*
> solution (lye)
>
> *To make the lye, dissolve 900 g (2 lb) of caustic soda in*
> *4½ litres (1 gal) of soft water.*

Over a low heat melt the fat. Add half the lye in small proportions, constantly stirring until well blended. Stir in the remaining lye, keeping the mixture at a gentle heat until it thickens to the consistency of thick pea soup.

When saponified, separation of the mixture has occurred, pour off the lye, add water to the mass, heat until dissolved, and again separate by a strong solution of common salt. (This part of the process is to purify the soap and can be omitted if a cruder product will suffice.) If the first separation does not readily occur, more lye should be added; the soap is insoluble in a strong caustic soda solution.

Remelt the soap in a double saucepan, at a gentle heat, until as much water as possible has been expelled. Pour into moulds, allow to harden, cut into bars and cure for several weeks.

Washing Machine Soap Powder

A biodegradable soap powder that is easy and simple to make, and has a gentle cleansing action on your clothes.

> 20 pbw crystallised washing   1 pbw basic soft soap, grated
> soda
>
> 5 pbw hard laundry soap,
> grated

Put the grated soap and half the washing soda in a pan, and melt the ingredients, stirring constantly, over a low heat. Care must be taken not to allow the mixture to reach boiling point.

Remove from heat and immediately stir in the remaining soda, until the whole mixture has become liquid and thoroughly blended. Keep stirring until it cools and thickens, then pour onto large, shallow metal trays to cool. Using a flat spatula, turn occasionally until the soap has completely cooled and is thoroughly broken up. Rub through a coarse sieve (wire netting over a timber frame is ideal); if you desire a fine soap powder, repeat the procedure twice more, each time through a finer sieve.

Store in large, lidded containers.

Soap Flakes
Ideal for hand washing delicate garments and baby clothes.

    9 parts grated soap   *1 part purified borax*

Put grated, unscented creamy white soap and borax in a double boiler and, stirring constantly, blend together over a gentle heat. Pour into moulds and allow to cool and harden.

To flake the soap, run it over an inverted carpenter's plane. Store in a lidded container.

Lavender Wool Wash
This soft soap is good for washing your blankets and woollens, and will leave them with just a hint of lavender fragrance.

    500 g (17¾ oz) grated, basic   *1 l (2 pt) boiling water*
    soft soap   *1 tbsp (1⅓ US tbsp) borax*
    4 tsp dried lavender

Put the dried herbs in a ceramic bowl and pour in the boiling water. Cover, steep overnight, then strain through muslin cloth.

Combine the grated soap, lavender water and borax in a saucepan. Bring to the boil, stirring constantly until well blended, remove from heat and cool. You will have a soft, white, jelly-like substance; store in a tightly sealed, wide-mouthed jar or other suitable container.

Wash woollens by hand, using ½ to 1 cup of the mixture, or 2 to 3 cups in the washing machine for blankets. Rinse well in clean water and then a final rinse of lavender, rosemary and wormwood. This will help to keep the moths at bay.

Prepare a herbal infusion using dried herbs in the following proportions:

    4 tsp lavender   *2 tsp wormwood*
    2 tsp rosemary

Put herbs in a ceramic bowl and pour in 600 ml (20 fl oz) of boiling water. Cover, infuse overnight, strain through muslin cloth and add to rinse water until it is sufficiently scented. Excess infusion can be used in the bath.

Washing Water Conditioner
Add this to your washing machine along with the soap powder.

It acts as a water conditioner and softener, and also helps to clean your clothes.

4 tbsp (5¹/₃ US tbsp) dried soapwort (root, leaves or stems)

1 tbsp (1¹/₃ US tbsp) dried lavender

1 dsp purified borax

60 ml (2 fl oz) methylated spirits

6 l (1¹/₂ gal) distilled water

Place the dried herbs in an enamel pan, add 2 litres (4¼ pt) of distilled water, bring to boil and then gently boil for 5 minutes. Remove from heat, stir in borax until dissolved, cover and infuse until cool. Strain through muslin cloth, squeezing all the liquid from the herbs, and add the remaining water and methylated spirits. Store in suitable tightly capped bottles.

Add one cup of liquid to wash water.

Eucalyptus Washing Liquid
This washing liquid is excellent for washing away dirt from all garments, especially grease from overalls.

1 kg (2¼ lb) homemade soap flakes

125 ml (4 fl oz) methylated spirits

20 ml (²/₃ fl oz) eucalyptus oil

Dissolve the soap flakes in methylated spirits, stand overnight, and then add eucalyptus oil. Do not be concerned if the flakes do not completely dissolve. Store in tightly capped bottles.

Add 1 tablespoon (1¹/₃ US tbsp) of the mixture for every 9 litres (2¼ gal) of water.

Prewash Conditioner
Spray onto perspiration stains and the like before washing the garment.

4 tbsp (5¹/₃ US tbsp) dried soapwort

1 l (2 pt) distilled water

vodka

Put the herb in an enamel saucepan and add distilled water. Bring to the boil and then gently boil for 30 minutes. Remove from heat, cover the pan, and allow to steep overnight. Strain through muslin cloth and add 10 ml (⅓ fl oz) of alcohol to every 500 ml (1 pt), or part thereof, of soapwort solution.

Store in pump spray bottles.

Herbal Fabric Wash

Use for washing delicate fabrics and woollens.

1 cup of bran	*1 l (2 pt) distilled water*
4 tbsp (5⅓ US tbsp) soapwort	

Boil the bran and soapwort together in 1 litre (2 pt) of distilled water in an enamel saucepan for 20 minutes. Strain through muslin cloth, squeezing all liquid from the ingredients.

Wash by hand in the solution, then rinse.

Stain-removing Soap

This soap can be used to remove stubborn stains from fabric.

1 handful of strawberry leaves	*900 g (2 lb) grated, basic hard soap*
1¼ l (2⅔ pt) white vinegar	*225 g (8 oz) precipitated chalk*

Place the strawberry leaves and vinegar in an enamel pan, bring to the boil, pour into a large wide-mouthed glass jar, seal, and leave where it will receive plenty of hot sunlight for two weeks.

Strain vinegar, squeezing all liquid from the herb, and add all ingredients to an enamel pan. Bring to the boil, and then continue boiling, stirring constantly, until the vinegar has evaporated. Pour mixture into moulds, allow to harden, and cut into cakes.

To use, wet stain with white vinegar and rub it with the soap. Leave the garment to dry in the sun before washing out.

Laundry Bleach

For those people who still find it necessary to use a bleach, this can be made using the same process as soap lye. However, the process is not carried out over a number of days; instead, pour water over the ash until sufficient liquid has been collected for your needs. Strain through muslin to remove particles of ash.

Use this lye for bleaching whites before washing. If you spread the bleached garments in the sun, stains will fade even more, making the washing easier.

Starch
Starch isn't used quite as much today. However, a strong starch can be easily made from old potatoes. Grate them coarsely under water; leave the mash to soak for twenty-four hours, strain off the dirty liquid, and spread the potato on trays to dry. You will be left with a powdered starch; store in an air-tight container.

Spray-on Starch
This can be used to spray on garments whilst ironing, and if perfumed with an aromatic herb will give your clothes a delightful, soft fragrance.

> 2 tsp dried lavender
> (or fragrant herb of your
> choice)
> 600 ml (20 fl oz) boiling
> water
>
> 10 ml (¹/₃ fl oz) vodka
> sufficient homemade starch
> ¹/₂ tsp gum tragacanth

Put the dried herb in a ceramic bowl and pour in the boiling water. Cover, steep overnight, and strain through muslin cloth, squeezing all the liquid from the herbs.

Warm the infusion and dissolve the gum tragacanth, then stir in sufficient powdered starch to make a viscous, creamy liquid. Add the vodka, stirring until completely blended. Store in a pump-spray bottle.

Lavender Ironing Spray
Spray on clothes when ironing; it will leave them with a fresh, soft scent.

> 2 tsp dried lavender
>
> 10 ml (¹/₃ fl oz) vodka

Prepare a herbal infusion from the lavender (see chapter 8). Add the vodka to the strained infusion, mix well, and store in a pump-spray bottle.

Lavender Washing Rinse
One or two cupfuls of the above infusion can be added to the final rinse of your washing machine cycle.

OTHER HOUSEHOLD SOAPS

Marine Soap
This is made by substituting coconut oil for other oils used in

making the basic soft soap. It has the advantage of forming a lather with salt water.

Follow the directions for basic hard soap, or use the following recipe:

30 pbw coconut oil	17½ pbw lye solution
3 pbw castor oil	

Mix the oils together in a large ceramic bowl or enamel pan, add the lye solution and stir for 5 minutes. Pour into moulds and keep in a warm spot for 24 hours, cut into bars, and then cure for several weeks.

The lye solution will become quite hot when the caustic soda is dissolved in water; set aside until lukewarm before adding to the oils.

Massage Soap

A soft soap that will lubricate the skin for smooth and comfortable stroking and kneading.

250 g (9 oz) grated pure coconut oil soap (use basic soft soap recipe)	1 tsp soya oil
	fragrant oil
5 g (⅙ oz) anhydrous lanolin	

Place grated soap in a pan with just enough water to cover. Melt over a medium heat, stir thoroughly, then add lanolin and oil. Continue to stir until the lanolin has melted and all the ingredients are well blended. Add a few drops of fragrant oil for scent and remove from heat. Pour liquid soap into large, wide-mouthed containers and allow to cool.

Pumice Soap

A good soap for removing ingrained stains and grease from hands.

350 g (12½ oz) caustic soda	2 kg (4½ lb) tallow
1 l (2 pt) soft water	powdered pumice stone
750 ml (25 fl oz) coconut oil	

Prepare as you would for basic hard soap. Stir in sufficient, sifted pumice to make a soft paste just before pouring into moulds. Cut into bars as soon as the mixture sets and allow to cure before using soap.

For vegans, the soap can be made using only coconut oil and the recipe for basic soft soap.

Sandsoap
Make in exactly the same way as pumice soap, replacing the powdered pumice with finely sifted, clean white sand.

Herbal Hand Cleanser
Although not a soap, this cleanser has been included because it will remove ingrained dirt and stains from hands.

30 ml (1 fl oz) olive oil	equal quantities of:
20 ml (²/₃ fl oz) avocado oil	dried sage
sugar	dried yarrow

Finely grind the herbs with a pestle and mortar and mix them together. Mix 1 tsp of the ground herbs and sufficient quantity of sugar with the oils to form a paste. Store in screw-top jars. Rub hands with paste until they are free of stains.

Liquid Antiseptic Soap
An ideal antiseptic soap for washing hands. Store in a pump-action bottle for easy use.

40 pbw olive oil	3 pbw tea tree oil
10 pbw thyme oil*	15 pbw potassium carbonate
7 pbw marigold oil*	alcohol and water, sufficient of each
(*see enfleurage, chapter 8)	

Prepare the same as dishwashing liquid, this chapter.

Blackwood Soap
This can be made by using the leaves of the blackwood wattle (*Acacia melanoxylon*), and will clean the dirtiest and greasiest hands. Add 2 handfuls of leaves to either warm or cold water and rub vigorously between your hands.

Carpet Soap
Use for removing dirt, stains and stubborn spots from carpets and mats.

112 g (4 oz) Fuller's earth	sufficient basic soft soap, grated
28 g (1 oz) spirits of turpentine	

> 224 g (8 oz) potassium
> carbonate (potash)

Put grated soap in a pan, with just enough water to cover, and melt over a medium heat, stirring constantly until well blended.

At the same time blend the other ingredients in a ceramic bowl, then add sufficient liquid soap, stirring continually, to make a stiff paste.

Store in a wide-mouthed container with a tight-fitting lid.

Saddle Soap

> 6 cups tallow
>
> 1 cup caustic soda

> 2½ cups soft water
>
> 1 cup glycerine

Heat the tallow to 54°C (130°F), stirring until completely melted. Dissolve the caustic soda in the water; allow it to cool to 35°C (95°F) and pour it slowly into the tallow, stirring for about 5 minutes. Stir in glycerine and pour into moulds to harden. After 24 hours cut into bars and cure for six weeks.

Dog Washing Soap

A good soap for washing the family pooch that will help control fleas.

> 5 pbw vaseline
>
> 4 pbw beeswax
>
> 5 pbw methylated spirits

> 15 pbw homemade laundry soap, grated
>
> fresh tansy or pennyroyal

Infuse one handful of fresh herb to every 200 g (7 oz) vaseline. Finely chop the herb; put the vaseline into a double saucepan and melt the vaseline over a medium heat. Add the herb and simmer for twenty minutes.

Stir in the wax and when melted add the grated soap, stirring until completely dissolved. Remove from heat and immediately stir in the alcohol. Pour into moulds to harden and cut into bars. This soap can be used immediately.

4 AROUND THE HOME

It was not that many years ago that herbs played an important role in household economies. They were used as medicines for people and livestock, strewn on floors to help eliminate odours and to control insect pests, to sweeten and purify musty air, and to clean, disinfect and polish.

Today, herbs can be used equally as well in the modern household, and are completely safe for humans and the environment. For each chemical product available in the supermarket, there is a natural alternative that is easy and simple to make. In this chapter I will be looking at housecleaning, eliminating unpleasant odours, household hints, pet care, insect pests and other natural or herbal based products.

GENERAL ADVICE

When making the various preparations for homemade toiletries, always used distilled water. Herbal infusions and decoctions also must always be prepared with distilled water. Household products, such as cleaners, polishes, air-freshener sprays, etc, can be made with tapwater, unless otherwise directed.

All you will need for making these recipes is basic kitchen equipment, and a few other odds and ends. However, before you start there are a number of precautions that should be noted, so as not to contaminate or mar the various preparations.

- Do not use aluminium, metal or non-stick pans for boiling or steeping herbs and flowers, or for the preparation of herbal recipes. Use only stainless steel or enamel pans for boiling and ceramic or glass pots for steeping
- All equipment must be kept scrupulously clean, and preferably used only for the preparation of herbal recipes. This safeguards them from contamination by foodstuffs and other foreign substances. Again, do not use metal or

aluminium, and always use a wooden spoon to stir, particularly when heating up ingredients

■ For toiletries, ointments, etc, sterilise containers and lids. Preferably use glass jars. Glass jars and metal lids should be thoroughly washed and rinsed clean, then placed in a saucepan of water, brought to the boil, and allowed to gently boil for ten minutes. Remove from heat and leave containers in hot water until ready for use

■ Adequately label everything you make — don't rely on memory

■ Cosmetic lotions and creams can have their keeping qualities extended by adding 8-12 drops of lemon juice, unless already specified, or 3-4 drops of tincture of benzoin (friar's balsam)

RECIPES

To simplify the recipes, the following abbreviations have been used in the text:

g	grams	*tbsp*	tablespoon
ml	millilitres	*dsp*	dessertspoon
l	litres	*tsp*	teaspoon

HOUSECLEANING

Making alternative products from scratch is cheaper, it cuts out the price mark-ups by all those middlemen and eliminates the cost of wasteful and expensive packaging and advertising. All these herbal products will make cleaning and polishing an aromatic delight, as well as being kinder to your skin and the environment.

Bathroom Cleanser

This can be used to clean baths, sinks and tiles, and is especially good where there is a build-up of grime or mould.

2 tbsp (2²/₃ US tbsp) dried soapwort

1 tbsp (1¹/₃ US tbsp) dried rosemary

1 cup homemade soap flakes

1 cake homemade laundry soap, grated

white vinegar, sufficient

75 g (2²/₃ oz) precipitated chalk (calcium carbonate)

Place the herbs in an enamel or stainless steel pan, add sufficient water to just cover, bring to the boil and reduce to a simmer for 30 minutes. Strain through muslin cloth, squeezing all the liquid from the herbs. Blend two parts herbal water to one part vinegar.

Add grated soap and soap flakes to a pan with just enough herbal liquid to cover. Melt over a medium heat, stirring constantly, then add the chalk and more of the herbal liquid to form a soft, thick paste. Store in wide-mouthed lidded containers.

Apply with a cloth, then rinse off.

Disinfectant Toilet Cleanser

2 tbsp (2²/3 US tbsp) dried soapwort

1 tbsp (1¹/3 US tbsp) dried thyme

1 tbsp (1¹/3 US tbsp) dried lemon verbena

white vinegar and baking soda, sufficient

Place all the herbs in an enamel or stainless steel saucepan, cover with water, bring to the boil and then simmer for 30 minutes. Strain through muslin cloth, squeezing all the liquid from the herbs. Blend one part vinegar to two parts herbal infusion, and then sufficient baking soda to form a paste. Store in a wide-mouthed jar.

Use this cleanser to remove stains from the inside of the toilet bowl.

Disinfectant Cleansing Liquid

Use this for daily cleaning of sinks, baths, toilets and kitchen benchtops, when all that is required is a wipe-over. It can also be used to clean and disinfect bathroom, kitchen and laundry wall and floor tiles and will remove surface mould from painted bathroom walls.

4 tbsp (5¹/3 US tbsp) dried soapwort

2 tbsp (2²/3 US tbsp) dried lemon verbena

1 tbsp (1¹/3 US tbsp) dried thyme

32 drops tea tree oil

40 ml (1¹/3 fl oz) alcohol

sufficient water to make 2 litres (4¹/4 pt)

Place the herbs in an enamel or stainless steel pan and cover with 1½ cups of water. Bring to the boil, then reduce to a simmer for 30 minutes. Remove from heat, cover and allow to infuse until cold. Strain through muslin cloth, squeezing all liquid from the herbs, and top up with water to make 2 litres (4¼ pt).

Mix tea tree oil with alcohol — either vodka or methylated spirits — and mix thoroughly with the herbal water. Use in a pump-spray bottle.

Herbal Disinfectant

Any of the following herbs can be used and are listed in descending order of their antiseptic power — lemon, tea tree, thyme, orange, bergamot, juniper, clove, lavender, niaouli, peppermint, rosemary, sandalwood and eucalyptus. The disinfectant can be made from the plant's essential oil or from an infusion of its leaves or flowers.

Essential Oil Method

12 drops of selected oil	2 l (4¼ pt) of tepid water
5 ml (1 tsp) alcohol (vodka or methylated spirits)	

Mix the essential oil with the alcohol (this aids dispersal), and add to the water. Do not use hot water, it will only evaporate the oil.

Infusion Method

4 tbsp (5⅓ US tbsp) dried selected herb	water sufficient to make 2 litres (4¼ pt)
40 ml (1⅓ fl oz) alcohol (vodka or methylated spirits)	

Place the herbs in an enamel pan, cover with 1½ cups of water, bring to the boil and simmer for 30 minutes. Remove from heat, cover and steep for two hours. Strain through muslin cloth, squeezing all liquid from the herbs, and add sufficient water to make 2 litres (4¼ pt). Add alcohol, mixing thoroughly, and store in a suitable tightly capped bottle.

Carpet Cleaner

This can be used for spot-cleaning carpets, tapestries and curtains, or, diluted, to lightly shampoo carpets and mats.

4 tbsp (5¹/₃ US tbsp) dried soapwort or	*distilled water, sufficient*
2 handfuls fresh herb	

Place soapwort in an enamel or stainless steel pan and add enough water to cover. Boil for 5 minutes, cool, and strain through muslin cloth.

Use neat for spot-cleaning or dilute 2 to 1 by volume, or more if needed, for shampooing.

Carpet Cleaning Soap — see chapter 3

Carpet Dry-Cleaning Paste
Use to dry clean natural fibre carpets and mats.
Mix sufficient water with Fuller's earth to make a paste and cover the soiled area. Allow to dry and then vacuum off.

Keep carpets smelling fresh by deodorising them with baking soda. Add the soda to a flour-sifter and sift finely over the area to be treated. Leave for 15 minutes then vacuum off. Add a few drops of your favourite fragrant oil to the vacuum cleaner and it will leave your carpets beautifully scented.

Vinyl and Ceramic Tile Cleaner

4 tbsp (5¹/₃ US tbsp) dried soapwort	*1 l (2 pt) white vinegar*
2 tbsp (2²/₃ US tbsp) dried lemon verbena	*water sufficient to make 1 litre (2 pt)*
12 drops tea tree oil	*5 ml (1 tsp) vodka or methylated spirits*
6 drops lemon oil (optional)	

Place the herbs in an enamel pan, cover with 1½ cups of water, bring to the boil and simmer for 30 minutes. Remove from heat, cover and steep until cool. Strain through muslin cloth, squeezing all liquid from the herbs, add sufficient water to make 1 litre (2 pt) and blend with white vinegar. Dissolve herbal oils in the alcohol, add to mixture and shake until completely mixed. Store in pump-spray bottles.

Spray onto floor surface to be cleaned and wipe over with a sponge mop.

Window and Glass Cleaner

Not only will this cleaner leave your windows and mirrors sparkling clean, it will also help to keep flies out when used on external windows or glass doors.

3 tsp dried lavender flowers	*8 drops lavender oil*
10 ml (1/3 fl oz) methylated spirits	*500 ml (16 fl oz) boiling water*

Place the lavender in a ceramic bowl, add boiling water, cover, steep overnight, then strain through muslin cloth. Dissolve the lavender oil in the methylated spirits, blend with the herbal infusion and drip through filter paper. Store in a pump-spray bottle.

Apply cleaner with a damp cloth, then buff off with wads of newspaper.

Grow Your Own Cleaning Sponge

An alternative to buying and using synthetic dishwashing sponges is to grow your own. A vegetable sponge, better known as a loofa *(Luffa aegyptiaca),* is a gourd that can be used, when dried, as a dishcloth, scrubbing pots and pans, wiping up spills, or cleaning human beings in the bath or shower.

Loofa gourds are grown like a zucchini and are a quick-growing annual that is more easily managed if given something to climb on, such as a back fence or trellis. When immature, under 15 cm (6″) long, they can be picked and eaten as a vegetable, tasting a little like zucchini or okra. They are best steam-cooked in a pressure cooker.

The fruit is ready to harvest for sponge-making when the stalk shrivels. Cut off the end, remove the seeds and hang in a sheltered spot, such as a verandah, so that the gourd will dry naturally in the sun. The fruit will turn yellow and wither and the skin will then flake off leaving the sponge-like fibres. Soak the sponge in clean water overnight, wash thoroughly in hot, soapy water, rinse clean, dry in the sun and it is ready for use.

Maintain your sponge and keep it in good condition by occasionally washing in warm water, to which has been added a little bicarb-soda. Thoroughly rinse in clean, warm water.

Discarded sponges can be added to the compost, and if you

have a bumper crop each year barter excess sponges with neighbours, or give them away as gifts.

Pot Scourer

The natural forerunner to steel wool and the synthetic scourers was horsetail stems *(Equisetum arvense)*. They have a fine sandpaper surface of silica crystals which will easily clean the dirtiest pots and pans.

To clean pots rub a handful of dried, leafless stems on their surface, then rinse to remove any residual green stains.

Horsetail can be easily grown in the home garden and will grow anywhere except in very alkaline soils. It is a perennial, fern-like, non-flowering plant that loves moisture and develops from spores, and is available from specialist herb nurseries.

Metal Cleaner and Polish

Horsetail makes an excellent cleaner for all types of metals, including pewter.

> *250 g (9 oz) fresh horsetail*   *denatured alcohol, sufficient*
> *1150 ml (38 fl oz) cold water*

Place the herb in an enamel or stainless steel pan, add the water and allow to infuse overnight. Boil for 15 minutes, remove from heat, allow to cool, strain through muslin, squeezing all liquid from the herb, and bottle. Add 5 ml (1 tsp) of alcohol to every 300-500 ml (10-16 fl oz) of liquid or part thereof.

Rub the metal pieces with a cloth soaked in the solution and leave to dry. Polish off with a soft cloth.

Furniture Polish

This simple, easy-to-make natural polish will leave furniture sweetly scented.

> *30 g (1 oz) beeswax*   *10 g (1/3 oz) very finely*
> *30 ml (1 fl oz) spirits of*   *ground sweet cicely seeds*
> *turpentine*

Melt the wax in a double saucepan over a very low heat. When completely liquid add the turpentine and sweet cicely seeds, stirring until all ingredients are well blended. Pour into a wide-mouthed jar and leave to cool before use.

Lemon Scented Furniture Cream

A soft cream polish suitable for all types of timber surfaces.

125 ml (4 fl oz) distilled
water

125 g (4½ oz) beeswax

25 g (1 oz) soap flakes (see
Laundry Needs, chapter 3)

500 ml (16 fl oz) spirits of
turpentine

1 handful dried lemon
verbena

Heat the distilled water in an enamel pan to boiling, add the lemon verbena, allow to cool, strain through muslin cloth, squeezing all liquid from the herbs, and stir in the soap flakes.

Melt the wax in a double pan over a medium heat. When completely liquid add the rest of the ingredients, stirring until well blended. Remove from heat, pour into a ceramic bowl and beat until cool and of a creamy texture. Store in a wide-mouthed jar with a tight-fitting lid.

Apply and polish off with a soft cloth.

Liquid Furniture Polish

A fragrant polish when all that is required is a quick wipe-over to remove dust and still leave timber surfaces shiny.

150 ml (5 fl oz) raw linseed
oil

150 ml (5 fl oz) white
vinegar

150 ml (5 fl oz) methylated
spirits

24 drops fragrant oil

Dissolve the fragrant oil in the methylated spirits, then mix together with the rest of the ingredients. Store in a spray-pump bottle and shake well before use. Spray sparingly and apply plenty of elbow grease when shining.

Furniture Rejuvenating Polish

Use to give new life to timber furniture that has been damaged with scratches.

150 ml (5 fl oz) linseed oil

150 ml (5 fl oz) spirits of
turpentine

25 g (1 oz) beeswax

Melt the beeswax in a double boiler over a low heat. When completely liquid stir in the linseed oil and turpentine until thoroughly blended. Store in a wide-mouthed jar with a tight-fitting lid.

Rub the polish well into the scratched parts, then polish off with a soft cloth. You will find that the scratches will almost be invisible.

Soft Cream Floor Polish

This polish can be used to polish both timber and linoleum surfaces, leather couches and shoes.

> 500 ml (16 fl oz) distilled water
>
> 500 ml (16 fl oz) spirits of turpentine
>
> 12 drops fragrant oil (optional)
>
> 75 g (2²/₃ oz) beeswax
>
> 25 g (1 oz) shredded laundry soap

Heat the distilled water in an enamel pan, add the shredded soap and stir constantly until completely dissolved.

Melt the wax in a double boiler over a medium heat until liquid. Add the rest of the ingredients, stirring to blend thoroughly, then pour into a ceramic bowl and beat until cool and of a creamy texture. Store in a wide-mouthed jar with a tight-fitting lid.

Apply lightly and polish off with a soft cloth or polishing pads.

Eucalyptus Floor Polish

An all-purpose disinfectant floor polish that can be used on timber, linoleum, vinyl and cork surfaces.

> 150 g (5¹/₃ oz) beeswax
>
> 250 g (9 oz) soap flakes
>
> 1 l (2 pt) boiling water
>
> 1 l (2 pt) spirits of turpentine
>
> 1 l (2 pt) raw linseed oil
>
> 1 handful eucalyptus leaves

Place the eucalyptus leaves in a ceramic bowl, add the boiling water, cover, infuse overnight and strain through muslin cloth.

Bring the herbal infusion to the boil, reduce to a simmer, add the soap flakes and stir until dissolved. Melt the wax in a double saucepan over a medium heat and when completely liquid stir in the herbal infusion. Remove from heat and when cool, but still liquid, add the rest of the ingredients, beating constantly with a wooden spoon until of a creamy texture. Store in a jar with a tight-fitting lid.

Shoe Polish

A neutral coloured shoe cream that will keep leather shoes supple and waterproof.

> 150 g (5⅓ oz) beeswax
>
> 50 g (1¾ oz) vaseline
>
> 50 g (1¾ oz) creamy white soap (see chapter 3)
>
> 1 l (2 pt) turpentine

Grate the soap into fine shavings, place in a ceramic bowl, add the turpentine and allow to stand for 24 hours. Blend with 1 litre (2 pt) of boiling water, stirring continually to dissolve any remaining soap.

Melt the beeswax and vaseline in a double boiler over a medium heat until completely liquid. Remove from heat, add the rest of the ingredients and stir constantly until cold. Store in a wide-mouthed open jar with a tight-fitting lid.

Coloured Shoe Polish

Coloured polish can be made by the addition of different earth pigments to the neutral shoe polish recipe.

The pigment can be any strongly coloured earth, sediment or clay. Boil it in water several times, each time in new water. Strain off and dry the remaining sediment in a warm spot. When completely dry, pulverise it to a fine powder using a mortar and pestle and store in containers with air-tight lids until needed.

To colour the polish, mix the powder with the water/turpentine/soap mixture until you get the colour you want. Alternatively, you can purchase different coloured earth oxides from a hardware or builder's supply store and use them instead.

Leather Rejuvenating Polish

This polish will restore leather furniture and shoes, and preserve it to prevent cracking.

> 550 ml (18 fl oz) raw linseed oil
>
> 275 ml (9 fl oz) white vinegar

Bring linseed to the boil over a medium heat, remove from heat source, allow to cool, and mix thoroughly with the vinegar. Store in tightly capped bottles until required.

Apply with a soft cloth, rubbing well into the leather. Keep changing to a clean area of the cloth as you go, then rub the leather with another soft, clean cloth until it shines.

HOUSEHOLD HINTS

Bathroom

Bath and Basin	*Basin Cleanser* or *Disinfectant Cleansing Liquid.*
	Blue/green copper stains can be removed by rubbing with a soft cloth moistened with vinegar. For old stains use *Bathroom Cleanser.*
	Bath stains can be removed with a paste of borax and lemon juice.
Ceramic Tiles	*Bathroom Cleanser.*
	Especially floor and shower recesses where there is a build-up of grime or mould.
	Wall Tiles — *Disinfectant Cleansing Liquid.*
	Bi-carb soda helps to remove mildew from small areas.
	White vinegar on a damp cloth for a quick wipe-over, or herbal disinfectant.
Mirrors (steaming up)	Rubbing with a cloth moistened with glycerine will prevent mirrors from steaming up while showering. Rubbing with soap and then polishing with a clean cloth will also work.
Mirrors (cleaning)	*Window and Glass Cleaner.*
Mould on Tiles	See Ceramic Tiles.
	Install ceiling fans or open windows to eliminate dampness that causes mould.
Shower Curtain	Scrub with white vinegar.
Shower Screens (glass)	*Window and Glass Cleaner.*

Shower Heads (blocked)

If shower heads become blocked, remove and boil in water containing ½ cup of vinegar.

Toilets

Clean the inside of bowl with *Disinfectant Toilet Cleanser*. Can also be used externally. Works especially well for badly stained bowls.

For a quick wipe-over use *Disinfectant Cleansing Liquid*.

Septic toilet bowls can be cleaned with white vinegar, it won't harm the microbes that break down the sewage in the tank.

Toilet bowl stains can be removed with paste of borax and lemon juice — do not use with a septic system.

Cars

Radiator Cleaner

Drain out the cooling system and flush it out with running water. Close drain outlets, add 200 g (7 oz) washing soda to the radiator, fill with clean water, and run the car for short distances only for 3 days. Drain the washing soda water from the system, flush thoroughly, and fill with clean water.

Do not use this cooling system if your car has an aluminium head or the cooling system is in poor condition.

Washing

Use a bucket and water only; do not leave the hose running; avoid the use of phosphate detergents; and wash the car on the grass if possible.

An effective alternative is to use the *Herbal Dishwashing Gel*, deleting the herbs if you wish. Or soft soap (from the chemist) dissolved in water. Wipe off with a chamois to remove any film.

Carpets

To remove dirt, stains and stubborn spots use the *Carpet Cleaning Soap*, chapter 3.

Spot clean carpets with the *Liquid Carpet Cleaner*. It can also be used for shampooing.

Dry-clean natural fibre carpets and mats with *Carpet Dry-Cleaning Paste*.

Deodorise with bi-carb soda — see Household.

Burn marks
: Rub with steel wool, working in a circular motion, then apply neat *Carpet Cleaner*.

Grease Spots
: Apply *Carpet Cleaning Soap* or *Carpet Cleaner* to the stain.

Ink
: Immediately apply a thick layer of salt to absorb the stain. Remove the salt.

Ballpoint ink
: Sponge with methylated spirits, then wash with *Carpet Cleaner*.

Wine, Beer, Coffee, Tea and Milk
: Immediately apply soda water to soak up the stain, then shampoo with *Carpet Cleaner,* if required.

Salt and bi-carb soda will absorb the stain. Apply thickly, vacuum when dry and shampoo if required.

Dog Urine Stains
: Sponge area with a diluted mixture of warm soapwort solution (see *Carpet Cleaner*) and ½ cupful of white vinegar.

Floors

Vinyl, Ceramic and Cork Tiles
: *Disinfectant Cleansing Liquid* or *Vinyl and Ceramic Floor Cleaner*.

For a quick wipe-over, mop with warm water and methylated spirits: 2 tbsp (2⅔ US tbsp) meths to a bucket of water. You can improve the disinfectant properties by infusing 4½ tbsp (6 US tbsp) dried lavender with the warm water first. Strain before using.

Slate	*Disinfectant Cleansing Liquid* or *Vinyl and Ceramic Floor Cleaner.*
Scuff Marks on Vinyl	Concentrated soapwort decoction, see chapter 8. Scrub or mop until marks disappear.
Unsealed Bricks	Remove stains by applying white vinegar. To bring them up to a beautiful shine apply the following polish:

½ cup lemon juice

1 cup olive oil

2½ ml (½ tsp) eucalyptus oil

2½ ml (½ tsp) natural turpentine

Thoroughly mix all ingredients and store in a tightly capped bottle. Label clearly.
Shake well before use.

Quarry Tiles	Clean with *Vinyl and Ceramic Tile Cleaner.* Grease stains can be removed with a neat soapwort decoction, see *Carpet Cleaner* or chapter 8.
Terrazzo	Clean with *Vinyl and Ceramic Tile Cleaner.* Stubborn stains can be removed with a cut lemon dipped in salt. Rub well into the area, leave 1 hour, then mop over with herbal cleaner.
Polishing Floors	*Soft Cream Floor Polish,* for timber and linoleum surfaces. Timber, vinyl, linoleum and cork surfaces — *Eucalyptus Floor Polish.*
Staining Timber Floors	Apply *Natural Timber Stain Finish,* see Household Products and Skills. To match or alter the colour of the stain,

boil the linseed oil first and add a little burnt sienna till the required colour is achieved. Then dilute with turpentine as directed.

Floor Wax Stripper

100 ml (3¹/₃ fl oz) white vinegar

230 ml (7¹/₂ fl oz) bi-carb soda

20 ml (²/₃ fl oz) eucalyptus oil

20 ml (²/₃ fl oz) methylated spirits

100 ml (3¹/₃ fl oz) household ammonia

4 l (1 gal) water

Dissolve the eucalyptus oil in the methylated spirits, then mix thoroughly with the rest of the ingredients. Store in a tightly capped, labelled bottle.

Furniture

General Furniture Polish

Liquid Furniture Polish.

Timber Surfaces

Lemon Scented Furniture Cream.

Polish and Dust Remover

Liquid Furniture Polish.

Rejuvenating Timber Furniture

To give new life to timber furniture that has been damaged with scratches, apply *Furniture Rejuvenating Polish.*

Leather Furniture

To restore leather furniture or to prevent it from cracking, apply *Leather Rejuvenating Polish.*

Stains and Scratches

Scratches
- Deep marks and scratches apply *Leather Rejuvenating Polish.*
- Light scratches — apply equal parts of olive oil and white vinegar. Rub well into the scratch with a soft, dry cloth.

Stains
- Water stains can be removed from

timber furniture by rubbing with a soft cloth dipped in camphorated oil.

- White heat marks — rub with camphorated oil.

- Leather — rub with eucalyptus oil, using a circular motion. Test a small section first in a spot that is not noticeable.
Apply *Saddle Soap*, chapter 3, rubbing well until the stain disappears.

- Chrome furniture — apply horsetail *Metal Cleaner and Polish*.
Can also be used for day-to-day cleaning and polishing.

- Cane and bamboo — use polish for unsealed bricks.
For cleaning or to remove light stains, make up a solution of warm water and salt. Brush on, then wipe off with a clean cloth.

- Aluminium — apply horsetail *Metal Cleaner and Polish*, rubbing in one direction only.

- Plastic furniture — use warm soapy water for general cleaning. Stains can be removed by rubbing with a cloth dipped in bi-carb soda.

- Fabric upholstery — day-to-day cleaning use diluted *Carpet Cleaner*, as directed.
For stubborn stains spot-clean with neat *Carpet Cleaner* or try *Carpet Dry-Cleaning Paste*.
Never over-wet. Dry liquid cleansers with a soft absorbent cloth using pressure only. Vacuum off dry-cleaning paste as directed.

Bruises on Timber Furniture	Wash indented places with warm water. Fold a sheet of brown paper several times, soak in water and place over the bruise. Apply a hot iron until all the water has evaporated from the paper. Repeat if necessary.
Stiff Drawers	Rub the inner and bottom edges of drawers with soap, beeswax, a candle or the basic *Furniture Polish*.

Jewellery

Gold Jewellery (including chains)	Place the ring or chain in a small glass bottle with a little warm water (never place two items in the one bottle). Add a little whiting (calcium carbonate — powdered chalk) and some grated pure soap (see chapter 3). Replace the cap and shake violently for about a minute. (Do not use this method for rings that have stone settings.)

The friction against the glass polishes the gold, while the other ingredients remove all traces of grease, grime or dirt. Works just as well on jewellery with intricate design work. |
| Gold Ornaments | Soak in a little warm soapy water, to which has been added a few drops of cloudy ammonia. Use a toothbrush to clean any intricate pattern.

Rinse, dip into alcohol and rub with a soft cloth. |
| Diamonds and Other Precious Stones | Soak in warm soapy water and brush with an old toothbrush. Rinse in water, dry, dip in alcohol and buff with a soft, clean cloth. |
| Silver | Bring ½ l (1 pt) of water to the boil, add 1 dsp of bi-carb soda, stir to dissolve, and a piece of silver foil. Place |

the articles in a large wire strainer and dip in and out of the boiling solution several times, shaking so that every part of the jewellery makes contact.

Rinse and dry with a soft, clean cloth.

To brighten silver rings soak overnight in ½ cup of water in which has been dissolved 3 tsp bi-carb soda. Rinse and polish with a soft, clean cloth. A parsley solution will also clean silver — see Kitchen, Metalware, this section.

Copper Jewellery

Soak in warm soapy water, then dip in a solution of 2 parts white vinegar to 1 part non-iodised salt. Rub dry with a soft, clean cloth.

Kitchen

Eliminating Odours

See Eliminating Kitchen and other Household Odours.

Chopping Boards

Use cold water to wash wooden cutting boards — hot water opens the grains and allows odours to sink in. Rub over with a handful of fresh lemon balm leaves or any of the mints.

Pastry Boards

To clean, scrape off left-over dough, sprinkle with salt and rub with a slightly dampened cloth or sponge.

Saucepans

■ Stubborn stains can be removed from aluminium pots by simmering in a strong solution of white vinegar for 20 minutes. Wash and use a wooden spoon to scrape.

■ To clean burnt saucepans, sprinkle with bi-carb soda and a dash of vinegar and bring to the boil. Wash when cool,

or,

simmer in a strong solution of cream

of tartar and sufficient water for 30 minutes.

- Stainless steel can be cleaned with a cloth dampened with vinegar. Rinse thoroughly.

- Copper pans — using a clean flannel, scour with a mixture of 2 tsp of vinegar to 1 tsp of salt. Wash in clean hot water.

- Enamel saucepans — if brown, can be cleaned by scouring with coarse salt, then washing in hot soda water.

- When boiling eggs, add a small slice of lemon to the water to prevent saucepans from discolouring.

- After cooking scrambled eggs, fill the pot with hot water and add a teaspoon of bi-carb soda. This will make the job of cleaning much easier.

- Even the dirtiest pots and pans will come clean with the use of a natural scourer, see *Pot Scourer.*

Metalware

The herbal *Metal Cleaner and Polish,* is ideal for cleaning all types of metal, including pewter. Other alternatives are:

- Aluminium — fine steel wool and plenty of soap.

- Brass — half a lemon dipped in salt.
  Mix vinegar and salt to a fine paste.
  mix common salt and sour milk.
  Finish off the above methods by a good rubbing with a soft cloth.
  Lacquered brass should be wiped over with a damp cloth or sponge.
  Lamp bases, fire tools and surrounds, etc, can be cleaned by rubbing in one

direction only with fine dry steel wool. Circular motions will scratch the metal.

- Bronze — wipe over with a little plant oil and polish with a soft cloth.
- Copper — mix 1 tbsp (1⅓ US tbsp) salt with a cup of white vinegar. Dip half a lemon in the solution and rub vigorously on the stain or the metal.
- Iron — soak in paraffin oil then rub with steel wool.
- Pewter — make a paste of fine, clean white sand and oil of tartar. Apply to the surface, then polish with a soft cloth.
- Silverware — to remove tarnish — boil up several handfuls of parsley and dip silver in this, or apply a little whiting (powdered chalk) dissolved in methylated spirits.

To polish and clean —

2 tbsp (2⅔ US tbsp) powdered alum

4 tbsp (5⅓ US tbsp) cream of tartar

4 tbsp (5⅓ US tbsp) whitening (finely sifted)

Mix all the ingredients thoroughly and store in an airtight container. Dampen a cloth with methylated spirits when applying.

To clean add 1½ tsp salt and 1½ tbsp (2 US tbsp) bi-carb soda to 1 litre (2 pt) of water. Bring to the boil and drop silver cutlery in. Boil for 3 minutes, then polish with a soft cloth. For badly tarnished silver add a piece of heavy duty aluminium foil — this will react with the soda to lift the tarnish.

To brighten silver in regular use, soak in a strong solution of borax and boiling water.

- Stainless steel — use soapy water or a cloth dampened with white vinegar.

Cutlery stains

- Egg stains on spoons — remove by rubbing with common salt.

- Knife blade stains — rub with the cut side of a raw potato, then wash with warm soapy water and polish.

- Bone handle stains — rub with a damp cloth dipped in salt.

For stubborn stains, dissolve a little salt in lemon juice and rub with a soft cloth, rinse in warm water and rub with a chamois.

Crockery (Plastic)

Stains can be removed by rubbing with a paste of equal parts bi-carb soda and coarse salt. Apply with a damp cloth and rinse clean.

Glass

- Deodorise jars and bottles — see Eliminating Kitchen and Other Household Odours.

- Cleaning glass decanters — add 1 dessertspoon of tea leaves, then wash well with warm soapy water. Rinse well with clean water.

- Sparkling glasses — rinse well in water to which has been added white vinegar.

- Beer glasses — just rinse and drain in cold water — never use soapy water.

- Pickle jars — if you rinse with cold

vinegar prior to bottling there is no need to sterilise the jars.

Electric Jugs and Kettles

■ To remove fur in a kettle, half fill with potato peel and top up with water. Boil for an hour, empty and rinse thoroughly. Top up with hot water if necessary.

■ Cleaning electric plastic jugs or kettles — add 1 cup of white vinegar, top up with water, boil for ten minutes, and rinse well.

Marble Flour Boards or Other Surfaces

■ To clean, first rinse off with clean water, then dissolve salt in lemon juice and rub on with a clean cloth.

Oven (cleaning)

■ Inside
Make a paste of bi-carb soda and water and spread it over the inside of the oven, then heat for 30 minutes. When the oven has cooled, brush the bi-carb off — it will remove all traces of burnt fat and grease.
To loosen foods that have become baked on, place 4 tbsp (5⅓ US tbsp) of bi-carb soda in the bottom of a shallow ovenproof baking dish (enamel, glass or ceramic) and add just enough water to cover the bottom of the dish. Heat the oven for 30 minutes, switch off, and leave the dish of bi-carb in overnight.

■ Outside — wipe the outside over with herbal *Disinfectant Cleansing Liquid*.

Stove

The outside of the stove and around hot-plates can be cleaned with herbal *Disinfectant Cleansing Liquid*.

For hard-to-remove burnt-on grime, apply a paste of bi-carb soda, leave on for 15 minutes, wipe off and then apply the herbal *Disinfectant Cleansing Liquid*. In nearly all circumstances the natural cleanser will remove even the most stubborn stains.

Coffee Pots and Teapots	To clean blackened coffee pots or tannin stained teapots, boil with bi-carb soda.
Laminex and Tile Bench Tops	*Disinfectant Cleansing Liquid*. Or, a paste of vinegar and bi-carb soda. Apply with a scourer and wipe off with a damp cloth.
Stainless Steel or Terrazzo Sinks	*Disinfectant Cleansing Liquid*.
Ceramic Wall Tiles	Stains and grease can be removed by wiping over with *Disinfectant Cleansing Liquid*, and rubbing into cracks and grooves to dissolve grease and grime.
Cooking Grease on Walls	Wipe over with *Disinfectant Cleansing Liquid* — apply pressure to stubborn spots.
Painted Woodwork (including cupboard doors)	Clean with *Disinfectant Cleansing Liquid*.
Stained Woodwork	Polished or stained cupboard doors and woodwork — see *Furniture Polish*, and Furniture, Stains and Scratches.
Refrigerator	■ To deodorise — see Eliminating Kitchen and Other Household Odours. Inside walls can also be wiped over with pure vanilla essence after cleaning.
	■ To clean, wipe outside and inside over with *Disinfectant Cleansing Liquid*. Finish off by wiping the inside with vanilla essence.

Drains	Avoid tipping cooking oil, fat, coffee grounds and tea leaves, etc, down the kitchen drains.
	To clear blocked drains, first try pouring a litre (2 pt) of hot, soapy water down, followed by plenty of hot tapwater. If this doesn't work, pour in half a cup of bi-carb soda followed by half a cup of vinegar, then stopper the drain and let the two substances react with each other for a while. Finally, unstopper the drain and run hot tapwater for a minute or so.
Washing the Dishes	To wash dishes by hand you can try a cake of the basic hard soap (see chapter 3) in an old-fashioned soap shaker — they are still available.
	See also *Herbal Dishwashing Liquid* and *Herbal Dishwashing Gel*.
	To eliminate that fishy smell sprinkle a little powdered mustard into washing water.
Chrome Taps and Other Fittings	Stained chrome can be cleaned with a cloth dipped in bi-carb soda. Polish with soft, clean cloth.

Laundry

Washing Machine and Dryer	To clean, wipe over with *Disinfectant Cleansing Liquid*.
Tub	As above.
Wall and Floor Tiles	To clean, see *Disinfectant Cleansing Liquid*.
Disinfectant	See *Herbal Disinfectant*.
Washing Clothes	▪ Always wash dyed articles separately for the first few washes, as they may run.
	▪ Before washing, always check clothes

for stains. Hot water will set the stains. Treat stains as directed.

■ Laundry Needs:

Pure Laundry Soap.

Washing Machine Soap Powder.

Soap Flakes.

Eucalyptus Washing Liquid.

Washing Water Conditioner.

Prewash Conditioner.

Stain-removing Soap.

Natural *Laundry Bleach.*

1 cup of lemon juice in 4½ litres (1 gal) of water can also be used as a bleach.

Starch.

Blankets and Woollens — see *Lavender Wool Wash.*

Delicate Fabrics and Woollens — see *Herbal Fabric Wash.*

Ironing

■ *Lavender Ironing Spray.*

■ To clean:
Clean the inside of a steam iron by filling with equal parts vinegar and water. Allow to steam a minute or so, switch off and leave for an hour. Wash out with clean water.

The face can be cleaned by applying bi-carb soda with a damp cloth, gently scrubbing until it is clean.

Starch can be removed from the face by sprinkling salt on a sheet of brown paper and moving the warm iron over it a few times.

Spots and Stains

The following spot and stain removing methods will work on natural fibres, though they may not always work on synthetic materials.

Spot Cleaning:

- apply neat soapwort decoction (see Decoction, chapter 8) and then rinse with clean water.

- eucalyptus oil is excellent for removing substances such as glue, tar, chewing gum and other unknown sticky substances from fabric.
Place a few drops of the oil on the stained area, leave for 2 minutes or until the oil has evaporated, and then wash.
Eucalyptus oil will also remove tar from shoes: apply the oil, leave for 2 minutes and then sponge with warm soapy water.
For persistent stains, prepare a soapwort decoction, diluted 2:1 (see chapter 8), then mix an equal quantity of the decoction with glycerine and soak for an hour. Rinse well in clean water.

Stains:

- Blood
Soak immediately in cold salted water for at least half an hour.
Mix cornmeal with water and rub onto the stain. Allow to dry, then brush away and wash as normal.

- Ballpoint ink — rub with methylated spirits then wash.

- Biro stains — see Ballpoint ink.

- Beer — use soapwort solution, diluted 2:1 (see Decoction, chapter 8).

- Cocoa — sponge with cold water, then rub with glycerine and leave for half an hour. Then wash with soap and water and rinse.

- Coffee
Mix an egg yolk with lukewarm water and rub onto the stain, then wash as normal.
To remove coffee from woollens, mix 10 ml (2 tsp) of glycerine, 5 ml (1 tsp) cloudy ammonia and 90 ml (3 fl oz) of water, apply to stain, rubbing gently. The mixture can be stored in a tightly capped bottle.

- Chocolate milk — as for cocoa.

- Chocolate — as for coffee.

- Curry — soak stain with methylated spirits, then wash as normal.

- Curtains
Spot-clean stains with soapwort decoction (see chapter 8), then wash. Dust and odours: soak curtains in vinegar, then wash.

- Egg — soak the stain in a cold solution of salt water.

- Fruit
Soak in milk before washing.
Rub immediately with powdered starch.
For dyed material that is not colour-fast, sponge with cold water followed by methylated spirits.
Dried-on stains should be moistened with alcohol first (brandy, whisky, etc) before washing.
Old stains can be rubbed with glycerine, left for 3 hours, then washed.

- Grass
  Sponge with methylated spirits and rinse with cold water.
  For heavy stains soak in eucalyptus oil for one hour, then wash.

- Grease
  Soak stained area in methylated spirits and wash with *Eucalyptus Washing Liquid.*
  To remove grease from carpets, cover with plenty of whitening (powdered chalk) and leave on for 24 hours. Vacuum up whitening, repeat procedure, then rub spots with a rag dipped in natural turpentine.

- Ink — apply a solution of lemon juice and salt, leave to soak and then wash in warm soapy water.

- Ink on carpet — cover thickly with salt immediately to absorb the stain. Vacuum off. Any residual stain may be rubbed with methylated spirits or neat soapwort decoction (see chapter 8).

- Lipstick — sponge with glycerine, leave for 30 minutes, sponge with eucalyptus oil, then wash in warm soapy water.

- Mildew — sponge with lemon juice and then place garment in the sun until the mildew spores have gone. Wash with soapwort fabric conditioner (see Herbal Fabric Wash).

- Milk — as for coffee on woollens.
  Or soak in cold water before washing.

- Nicotine — sponge with eucalyptus oil.

- Oil — see Grease.

- Perspiration
  Soak garments in a solution of 2 tbsp
  (2⅔ US tbsp) bi-carb soda and 4½
  litres (1 gal) of cold water. Leave for
  one hour before washing.
  Old stains can be sponged with
  vinegar before washing.

- Rust — soak in a strong solution of
  lemon juice and salt. Leave in the sun
  to dry, then wash.

- Scorch marks — treat as for rust.

- Tea — treat as for cocoa.

- Wax — allow to harden, scrape off
  as much as possible, and then remove
  the rest with methylated spirits.

- Wine — cover immediately with salt,
  and rinse in cold water before
  washing.
  On carpets — pour a little mineral
  water on the stain immediately, leave
  for a few minutes, then blot up with
  a clean towel.

Dry Cleaning

Mix Fuller's earth with water to form
a paste. Cover the soiled area, allow to
dry, then brush off (out of doors) or
vacuum. This method is suitable for
smooth fabrics, carpets and mats.

Walls, Windows and Glass

Walls — Painted   Clean with *Disinfectant Cleansing Liquid*.

Mould on walls — as above.

Grease on walls — as above.

Wallpaper   Rub with stale bread or a damp cloth
wrung out in herbal disinfectant. Be

careful not to damage the paper and use only light pressure.

Vinyl wallpaper can be cleaned with *Disinfectant Cleansing Liquid*, or a warm cloth wrung out in vinegar.

Fabric wall coverings — dab with a paste of whitening, leave for an hour or so then gently brush off.

Windows	Clean with *Window and Glass Cleaner*.
Glass	As for windows.

Miscellaneous

Paint	To help remove the smell, add a small amount of vanilla essence to the paint. Leave a bucket of water in freshly painted rooms.
Shoes	*Polish*, see *Shoe Polish*.
	Patent leather — rub with vaseline.
	Stains on suede shoes — cover with cornstarch and brush off after a few minutes, then hold over a steaming kettle and give it a good brushing with a suede brush.
Sticky Marks left by Tape or Price Tags	Wipe gently with methylated spirits and a few drops of eucalyptus oil. To remove tape from vinyl surfaces without leaving a mark, moisten well with eucalyptus oil, leave several minutes and lift off. You may need to repeat this a few times — then wipe over with methylated spirits and eucalyptus oil.

HOUSEHOLD PRODUCTS AND SKILLS

Herbs and other natural ingredients are completely safe, unlike many chemical-based products. Plants provide a wealth of natural, harmless extracts that can be used together with many other safe substances to provide alternative paint, dyes, stains, etc.

Household Paint The following paint can be coloured using dried earth pigments (see *Coloured Shoe Polish*) and is suitable for both inside and outside surfaces.

> *sour milk, sufficient* *slaked (hydrated) lime, sufficient (from the hardware store)*

Mix sufficient lime and sour milk until they have neutralised each other. Test with litmus paper: if the paper turns red add more lime, and if it turns blue add more sour milk.

To colour, mix powdered pigment with the paint until you get the shade you want.

Slaked lime can be made from powdered lime: add sufficient lime to a plastic bucket, cover with about 12 cm (5″) of boiling water, and stir in briskly till thoroughly slaked.

Whitewash A simple-to-make paint that will withstand rain and other weather conditions without coming off. It can be used on external surfaces, including galvanised iron.

> *3 kg (6²/₃ lb) tallow* *3 kg (6²/₃ lb) dry, powdered lime*

Place the tallow in a large bucket so that it completely covers its base, add the lime and then enough water to cover the ingredients by approximately 10 cm (4″). When the heat from the lime has melted the tallow stir the mixture thoroughly until the ingredients have dissolved and are well blended. Store in a container with tightly fitting lid.

Apply to dry surfaces with a large brush.

This paint will dry to an off-white and can be coloured by the addition of powdered earth pigments.

Filler for Fine Cracks before Painting Mix a little flour to some of the paint to be used to form a soft putty. Rub into cracks and gently wipe off surplus.

Gloss Whitewash for Timber Surfaces To render the basic whitewash more durable for timber surfaces, and to give it a bright gloss, add 500 ml (1 pt) of milk to every 4½ litres (1 gal) of paint just prior to application.

Natural Timber Stain Finish   External and internal timber surfaces can be sealed and preserved with warm linseed oil and will still retain their natural appearance. The first coat should be thinned 20 per cent with turpentine; the second coat with 10 per cent turpentine, and the last coat should be straight linseed oil. Allow two weeks between each coat.

Natural Dyes

In chapter 1 — Rags and Old Clothes — I discussed various ways in which old clothes and scraps of material can be recycled. All of these projects can be given even more life and an individual uniqueness by dyeing them.

Natural dyes from herbs and our own native plants are unsurpassed for richness and subtlety of colour. Although they take best on wool and silk, cotton and linen can be dyed with reasonable success; however, natural dyes will not colour synthetics except rayon.

The process is simple and is more or less an extension of making a herbal infusion. However, before you start you will need to prepare the fabric so that it will receive the dye; this involves scouring for wool and mordanting for all materials.

When treating wool and silk avoid sudden temperature changes such as lifting from a hot bath and then placing it on a cold surface. Always handle gently and dry away from direct heat; likewise for cotton and linen fabrics.

Scouring   Soak wool for several hours in warm water (about 50°C/122°F) with 1 tbsp (1⅓ US tbsp) of liquid detergent added (see *Herbal Dishwashing Liquid*, chapter 3). This is necessary to remove all traces of oil from the wool. Remove the wool, gently squeezing all the liquid from it, then repeat the process in clean water. Give a final rinse in warm water to which has been added about 50 ml (1⅔ fl oz) of white vinegar.

Follow the procedure for silk, with water at 90°C (194°F).

Mordanting   Most fabrics, especially wool, are naturally water repellant and many dyes will not take unless the fabric is treated with a mordant first. White vinegar is a readily available mordant; however, to get a greater range of colours you may need to use substances like alum, cream of tartar or acetic acid.

To treat your fabric, first dissolve the mordant in a little hot water, then stir it into 20 litres (5¼ gal) of water at 50°C (122°F). Thoroughly wet your wool or cloth, submerge it in the mordant, slowly bring to the boil over an hour, and then reduce to a simmer at 82-93°C (180°-200°F) for a further hour. Then rinse and dye immediately in a prepared dye bath.

Silk is treated differently and should be immersed at 60°C (140°F) and allowed to steep for 24 hours.

Mordants
Dissolve the following quantities of mordant to every 20 litres (5¼ gal) of water:

Alum — 25 g (1 oz)	Gives bright, clear colours.
Copper (sulphate) — 15 g (½ oz)	Handle with care. Combined with 250 ml (8 fl oz) of vinegar gives a blue-green tint to colours.
Cream of Tartar — 20 g (¾ oz)	Combine with alum for bright colours.
Iron (ferrous sulphate) — 5 g (⅙ oz)	Dulls and deepens colours. Use alum first in dye bath, simmer for 45 minutes, remove fabric, add iron, replace fabric and simmer for a further 30 minutes.
Vinegar — 275 ml (9 fl oz)	Will fix most dyes, but colours won't be as strong.

These quantities are based on using 450 g (1 lb) of dry wool.

Dye Bath Chop up selected plant material, wrap loosely in a square of muslin and secure with a piece of string, place in an enamel pan with 20 litres (5¼ gal) of soft, tepid water and allow to steep overnight. Simmer for 1-3 hours at about 90°C (194°F) until the desired colour is obtained, remove the herbs, cool to hand heat and gently add the fabric. Take one hour to return to a simmer, then simmer for a further hour. Allow to cool to hand heat, remove fabric, rinse in warm, tepid and then cold water and hang to dry.

Silk should be left to cool overnight, then rinsed in warm, tepid and cold water.

Herb Dye Chart The following list of herbs is by no means exhaustive and only serves as a guide to colours available from the more common plants. Try experimenting with different herbs, garden plants and Australian native plants and different mordants. You'll be surprised at the variety and richness of colours you can obtain. Reduce the quantity of the dye bath and test swatches of different materials.

HERB	PART USED	MORDANT	COLOUR
Agrimony	flowering tops	alum	butter yellow
Alkanet	root	acetic acid	soft pink-brown
Blackberry	young shoots	alum	creamy fawn
Bracken	young shoots	alum	yellowish green
Comfrey	fresh green plant	alum	yellow
Chamomile, dyer's	flowers	alum & cream of tartar	bright yellow
		copper or acetic acid	olive
Elder	leaves	alum & cream of tartar	greenish yellow
		copper or acetic acid	olive
		as above, plus a pinch of iron	grey green
Elder	berries	alum or salt	purple
Horsetail	fresh stems	alum	creamy yellow
Juniper	fresh crushed berries	alum	strong yellow
	crushed dried berries	alum & cream of tartar	olive-brown
Lady's bedstraw	roots	alum	coral pink
Madder	roots	alum or cream of tartar	rich tomato red

HERB	PART USED	MORDANT	COLOUR
Marigold	petals	alum or cream of tartar	pale yellow
Nettle	whole plant	copper	soft grey-green
Onion	skins	alum or cream of tartar	orange
		copper or acetic acid	deep brassy yellow
Parsley	fresh leaves & stems	alum	cream
Privet	leaves, young shoots	alum	strong yellow
	berries	alum	bluish-green
Safflower	flowers	alum	beige
St John's wort	flowers	alum	beige
Sorrel	whole plant	alum	greyish yellow
	roots	alum	soft pink
Tansy	flowering tops	alum	mustard yellow
Turmeric	powdered root	alum	gold-orange
Walnut	leaves	no mordant	creamy fawn
	green husks & shells	no mordant	shades of brown
Wild marjoram	leaves	alum	violet

Equipment Needed for Dyeing

wooden dowels for hanging skeins of wool
muslin cloth
rubber gloves
pestle and mortar for grinding herbs
plastic buckets for rinsing
enamel or stainless pan for dye bath
thermometer and scales

Remember, no two batches of herbal dyes will be the same: colour will always depend upon plant variety, sunlight it has received when growing, mordant used and the strength of the dye bath and immersion time.

Herbal Inks

Writing ink can be made from plants such as inkbush, mulberries or nightshade (*Solanum nigrum*) berries, which is not Deadly Nightshade. They can then be scented with fragrant herbs that will give your personal letters a delicate, intangible fragrance the moment they are opened.

Herbs to use are those that have a strong fragrance, such as lavender, rosemary, rose petal, lemon verbena or hyssop. Use herbs dried for the best effect.

Basic Ink

Crush the berries or fruit of any of the above three plants and squeeze out all the juice. Use rubber gloves to avoid staining hands and an apron to protect clothing.

Gently warm the juice and add one teaspoon of arrowroot to every 80 ml (2⅔ fl oz) to thicken it. Stir continuously until the mixture thickens and loses its cloudy appearance.

Lavendar Ink

> 15 g (½ oz) dried lavender   125 ml (4 fl oz) basic ink
> flowers
> 6 tbsp (8 US tbsp) water

Crush the lavender and put it into a saucepan with the water. Bring to the boil and simmer for about 30 minutes or until you are left with 40 ml (1⅓ fl oz) of liquid. Strain through muslin cloth, squeezing all liquid from the herbs, and then blend with the basic ink.

If the ink is not thick enough, gently warm and adjust by adding a little more arrowroot.

Lemon Verbena Ink

> 125 ml (4 fl oz) water   125 ml (4 fl oz) basic ink
> 25 g (1 oz) dried lemon verbena

Prepare the same as lavender ink, reducing the herbal water to 40 ml (1⅓ fl oz).

Hyssop Ink

> 25 g (1 oz) dried hyssop   125 ml (4 fl oz) basic ink
>
> 125 ml (4 fl oz) water

Prepare the same as lemon verbena ink, again reducing the herbal water to 40 ml (1⅓ fl oz).

Alternative method for fragrant ink

> 25 g (1 oz) dried fragrant   125 ml (4 fl oz) basic ink
> herb
>
> 100 ml (3⅓ fl oz) water

Prepare the same as lavender ink, reducing herbal water to 20 ml (⅔ fl oz) and then mixing with basic ink. This gives a far stronger fragrance.

Red Ink

> 1 cup of field poppy petals   water and alcohol (sufficient)

Place the flower petals in a ceramic bowl, add just enough boiling water to cover the petals, cover, steep overnight and strain through muslin cloth, squeezing all the liquid from the petals. Add 15 per cent alcohol to preserve the ink and bottle.

For the alcohol use vodka or methylated spirits. The latter will have a slight odour typical of a denatured spirit, which can be eliminated by adding a few drops of pure fragrant oil.

Various coloured inks can be made from this recipe by choosing the desired colour from those herbs listed in the herb 'dye chart'.

Glue

Glue can be made from any tree that gives reasonable quantities of resin, and will usually be coloured according to the resin's pigment. However, a clear glue can be made from the species of wattle, *Acacia senegal*. This is the gum traditionally used to make gum arabic.

Clear Glue

> small blob of gum   100 ml (3⅓ fl oz) boiling
>   water

Place the gum in a piece of muslin cloth and suspend this in the hot water until it dissolves. Any impurities will remain in the cloth, giving you a clear glue that can be used to stick paper together.

If the glue is not thick enough, adjust by using a large blob of gum.

An alternative method is to buy the powdered gum arabic and dissolve this in warm water until you have a thick solution.

China Cement

This will give you a white cement that is ideal for repairing broken china and other ceramic objects.

Prepare as you would the clear glue, then stir in sufficient plaster of Paris to make a thick paste. Apply with a brush to broken edges, and then press them together.

Candlemaking

Candles provide emergency lighting, and they have other uses in the home. Scented with fragrant oil they will eliminate cooking odours, freshen and delicately perfume stale air, leave bathrooms wonderfully aromatic, and repel insects.

Making candles is fun, simple and easy, and can be an inexpensive hobby that could earn you extra dollars at a craft market. Most candles are made from a mixture of paraffin wax and stearin (stearic acid), available from craft shops and hardware stores. However, I prefer to use beeswax: it is a high quality natural ingredient that gives good light, colour and texture. Specialty shops sell beeswax, but it is very expensive; you can, however, buy it direct from a beekeeper in your area, at a cost quite substantially less.

Wicks can be obtained from craft shops. As a general guide, the bigger the candle the bigger the wick. If it is too small, the melted wax floods the flame and drips down the candle.

Moulds can be made from almost anything, their shape and size depending upon your personal preferences. Make cylinders or odd shapes from cardboard held in place with masking tape. Use everyday items like toilet rolls, milk cartons, yoghurt containers or soft drink cans. For fragrant candles around the house I cast the wax directly into small, attractive ceramic bowls or glass jars.

When the wax cools it shrinks, so candles can be easily removed from moulds.

Adding the Wick

It is far easier to position the wick before adding wax to the mould. Poke a small hole in the bottom of the mould, push the wick through to the length required, knot it, seal the hole with masking tape, and then secure the other end to a pencil laid across the top of the mould. Position the wick so that it is running exactly through the centre of the mould.

When casting into permanent moulds, such as ceramic bowls, secure the bottom of the wick with plasticine and the top end as before. Remove the plasticine once the wax cools and shrinks.

Eucalyptus Candles

One of these aromatic candles burning in a room will help to repel mosquitoes and blowflies, as well as making the room smell clean and fresh.

> *beeswax (sufficient)*   *eucalyptus oil*

Melt the wax in a double saucepan over a medium heat so that it does not burn. When completely liquid add 20 ml (⅔ fl oz) of eucalyptus oil to every half litre (16 fl oz) of wax. Pour into moulds and allow to harden — 3 to 4 hours — before using. Trim the wick to 1 cm (⅜″) and smooth the sides with an old nylon stocking.

Fragrant Candles

Pure essential herb and flower oils (not alcohol based) can be added to the liquid wax just before pouring it into the mould. Add them a drop at a time until the wax is sufficiently scented. Homemade fragrant oils (see chapter 8) can also be used, adding about 5 ml (1 tsp) to every 100 g (3½ oz) of wax. The more oil added the stronger the scent.

Usually I use pure essential oils when making fragrant candles, and add 20 to 30 drops for every 100 g (3½ oz) of wax.

Herbal oils that are suitable for fragrant candles are: bergamot, geranium (all varieties), germander, jasmine, lavender, lemon oils (all varieties), lemon thyme, mint (all varieties), rose, rosemary and sweet violet. Basil and pennyroyal will help to deter flies, and hyssop adds a clean, fresh scent to bathrooms and toilets.

Paraffin Wax Candles
If you have to use paraffin wax, prepare as you would for beeswax. If you are going to add stearin, melt it in a separate pan and add it at the rate of one part to 10 parts of paraffin wax just before pouring it into the mould.

HERBAL FRAGRANCE IN THE HOME

The refreshing and aromatic properties of herbs and flowers can be used in many different ways to keep your home fragrantly fresh. They will remove stale and other unwanted odours, freshen up cupboards and drawers, keep shoes smelling clean and fresh, add a soothing drift of scent to the bedroom and give your linen and clothes a touch of fragrant luxury.

Room Freshener Spray
Use this to eliminate bathroom odours, freshen a sick room, or anywhere else that an antiseptic spray is needed. Choose from any of the following herbs (listed in descending order of their antiseptic powers): thyme, orange flower, bergamot, juniper, clove, lavender, niaouli, peppermint, rosemary, sandalwood, eucalyptus.

> *24-30 drops of essential oil* *500 ml (1 pt) distilled water*
> *5 ml (1 tsp) vodka*

Dissolve essential oil in alcohol, add to distilled water and store in a pump-spray bottle. Shake well to mix and use on a fine mist setting.
Or

> *6-8 tsp dried herb* *500 ml (1 pt) distilled water*
> *10 ml (2 tsp) vodka*

Place the herbs in an enamel saucepan, add the water, bring to the boil and reduce to a simmer for five minutes. Remove from heat, cover, steep overnight and strain through muslin cloth, squeezing all liquid from the herbs. Stir in alcohol until well blended and store in a pump-spray bottle.

Distilled Fragrant Water
A very strongly scented spray can be made by any of the various methods described in chapter 8. Use any of the herbs already recommended, or any other fragrant herb or flower where antiseptic qualities are not required.

Air Purification Spray
This is a very strong antiseptic spray to use in a sick room and to help combat the spread of germs throughout the house.

25 drops lemon oil	5 drops peppermint oil
25 drops lavender oil	5 drops eucalyptus oil
15 drops thyme oil	50 ml (1²/₃ fl oz) vodka
15 drops clove oil	500 ml (1 pt) distilled water
10 drops tea tree oil	

Dissolve the essential oils in the alcohol and blend with the distilled water. Store in a pump-spray bottle and use as required.

This spray can also be made using just the tea tree and eucalyptus oils only and replacing the rest of the ingredients with dried herbs. Use dried lemon verbena instead of lemon oil and then prepare the herbs as described for Room Freshener Spray method 2. Blend the oils with the alcohol and mix with the herbal water.

Essential Oils

Room Fresheners

- add a few drops of fragrant oil to a shallow dish of warm water set on a sunny window sill or radiator. As it evaporates its aroma will fill the room
- in a sickroom or bathroom, moisten a sponge with boiling water and add a few drops of essential oil. Place the sponge in a dish in the room, and moisten it with boiling water twice a day, refreshing it with a few drops of oil twice a week. Use lavender or peppermint oil in a sickroom as protection against bacteria
- one or two drops of oil on a warm light bulb will quickly fill the room with its fragrance. Place a few drops of one or more oils in a light bulb ring, turn on the light and a gentle fragrance will waft around your home. Do not use oils containing alcohol.

    Burning essential oils will kill airborne bacteria and fungi. Try thyme, lavender, pine or eucalyptus for their fresh fragrance.

    Light rings can be used on both ceiling lights and lamps

and are available by mail order from The Fragrant Gardens, Portsmouth Rd, Erina, NSW 2250.

- a few drops of essential oil in the vacuum cleaner bag leaves a delightful fragrance as you clean the house
- homemade scented candles will fill your room with a warm, comforting fragrance, and are ideal to burn in dry areas where there is little humidity
- a simmering pot is a small ceramic vessel with a candle inside. A small saucer containing a mixture of fragrant oil and water sits on top and the burning candle releases its fragrant vapours. Simmering pots are available from most herb, craft and gift shops and are designed to sit on a dining table or in the lounge room.

  Add about 10 drops of fragrant oil to one cup of boiling water, preheat the saucer with the candle burning, and then three-quarter fill it with the fragrant water, topping up when required
- terracotta scent pots are another delightful way to scent rooms or cupboards with essential oil. Hang them in an appropriate place (the clothes rail in your wardrobe is perfect) add 6 drops of oil to start, then 1-2 drops every week. The terracotta holds the fragrance and gently releases it into the air.

  Like the simmering pots, they are also available from herb and craft shops.

Car freshener

A few drops of essential oil sprinkled on a tissue and placed in the air vent of your car will keep it fragrant fresh as you drive. Citrus oils will refresh stale air and basil or peppermint will help you to remain alert over long distances.

Studying

Long hours of study are taxing to both the mind and body. To help you cope and remain alert add a drop of essential oil to a page in each book you are using.

Try any of the following:

Basil	to clear your head
Bergamot	to bring freshness

Cardamon	for reducing mental fatigue
Lavender	ideal for physical and mental tension
Rose	will lift your spirits
Tangerine	energising

Pot-Pourri

The traditional way to capture the wonderful fragrances of a summer herb garden all year round. Fragrant mixes can be displayed in glass lidded bowls to release their aromatics whenever required, pressed into recycled paper to gently perfume drawers placed in bags and hung in cupboards, closets and wardrobes, or stuffed into pillows for blissful sleep.

To perfume your home with the old fashioned magic of yesteryear is simple and uncomplicated. The following recipes will suit most situations and if cared for will last for many years. They are based on the dry-mix method for ease of preparation and success.

How to blend a dry pot-pourri

In a large ceramic bowl mix together well all the dry ingredients, spices and fixative. Use your hands to ensure a good mix.

Add the appropriate essential or fragrant oil one drop at a time. Mix in well after each addition and test for scent.

Put the pot-pourri in glass jars that have air-tight lids, or in well sealed plastic bags, and leave to cure in a dry, dark spot for 6 weeks. Be sure to give the mix a good shake every other day.

Drying flowers and herbs

Pick twice as many fresh flowers and/or herb leaves as called for in the recipe, plus 10 per cent. Dry them until they are crisp, yet still maintain most of their colour.

To dry flowers and leaves simply tie them in separate bunches and hang upside down in a dry, airy place, or spread them thinly on net-covered drying trays and keep in a warm, well ventilated place. Usually, herbs will take from 4-12 days to dry, sometimes longer.

Leaves	are dry when brittle, but will not shatter.

Flowers are ready when petals feel dry and slightly crisp.

Pot-Pourri Blends

Lemon Pot-Pourri

1 cup dried lemon verbena leaves

1 cup dried lemon balm leaves

1 cup chamomile flowers

1 cup marigold petals

½ cup orrisroot powder

6 drops lemon verbena oil

dried peel of one lemon, powdered

To dry citrus peel, thinly pare the skin, making sure that no pith is left attached to it, and place in the sun until completely dried. Stud the peel with cloves and dry in a warm oven (100°C/ 210°F) until crisp. Remove and powder in a pestle and mortar or blender.

Aromatic Insect Repellant Blend

2 cups dried lavender flowers

1 cup dried rosemary

1 cup dried southernwood

½ cup dried spearmint

½ cup dried santolina

¼ cup dried pennyroyal

¼ cup dried tansy

¼ cup cedarwood chips

3 tbsp (4 US tbsp) orris root powder

Australian Country Pot-Pourri

2 cups dried rose petals

½ cup freesias

½ cup acacia flowers

1 cup eucalyptus leaves (torn and crushed)

½ cup dried lemon scented geranium leaves

2 cups dried lavender

½ cup dried rosemary leaves

½ cup orrisroot powder

1 tbsp (1⅓ US tbsp) ground dried lemon peel

3 drops rose oil

2½ drops lavender oil

1½ drops eucalyptus oil

Lavender Pot-Pourri

3 cups dried lavender buds

½ cup dried jasmine flowers

2 tbsp (2²/₃ US tbsp) ground
dried lemon peel

2 tbsp (2²/₃ US tbsp) dried
basil leaves

2 tbsp (2²/₃ US tbsp) dried
rosemary leaves

6 drops lavender oil

4 tbsp (5¹/₃ US tbsp) dried
spearmint leaves

4 tbsp (5¹/₃ US tbsp) orrisroot
powder

1 tbsp (1¹/₃ US tbsp) gum
benzoin

Rose Pot-Pourri

4 cups dried red rose petal
and buds

2 cups dried rose scented
geranium leaves

3 cups dried lavender buds

2 tbsp (2²/₃ US tbsp) freshly
ground cinnamon

2 tbsp (2²/₃ US tbsp) freshly
ground allspice

¼ cup freshly ground cloves

1 tbsp (1¹/₃ US tbsp) coarsely
ground mace

⅓ cup orrisroot powder

6 drops rose oil

3 drops lavender oil .

Using your pot-pourri

Pot-pourri may look very attractive displayed in an open bowl,
but its fragrance will not last long. Store in closed containers
away from direct sunlight. (Tightly sealed glass jars will still allow
you to enjoy the beauty of the mix.) Open the container each
day for about one hour to allow the fragrance to waft around
the room. You will find that it will last for hours.

Refresh the mix once a week with the appropriate fragrant
or essential oil. One or two drops should be sufficient, but always
test the scent before adding more. Too much oil, or even the
wrong one, will ruin the fragrant nose and character of your
composition. Never let the pot-pourri dry out, try and avoid damp
and humid atmospheres, and never add alcohol, perfume or
scented toilet waters; these will only cause the fragrance to rapidly
deteriorate.

Winter Pot-Pourri

This mix will keep rooms smelling sweet during those cold winter months. Place the pot-pourri in a metal bowl on top of a hot radiator; its scent will soon waft throughout your house.

equal parts of:	8-9 drops of essential oil of:
ground cloves	cloves
ground cinnamon	rose
lavender	lemon verbena
orrisroot and gum benzoin	

Mix all ingredients thoroughly together and store in a tightly sealed glass jar for a week. It is then ready to use.

Herbal Incense

This pot-pourri needs to mature for 6-12 months to develop its aromatic properties fully, but is well worth the wait, especially when burnt as incense.

½ cup freshly crushed cinnamon	½ cup dried minced lemon verbena
½ cup crushed allspice	½ cup dried minced lavender leaves
½ cup crushed nutmeg	
½ cup minced vanilla pod	½ cup dried minced rosemary leaves
½ cup cedarwood raspings	
½ cup powdered dried lemon peel, (see Lemon Pot-Pourri)	15 ml (½ fl oz) rose oil

Prepare using the dry pot-pourri method, crushing larger pieces in a pestle and mortar. Place in a glass jar, with a tight-fitting lid, and shake every other day for the first two months, then leave to mature for at least another four months. The longer it's left the stronger it will be.

To use, place about 2 tbsp (3 US tbsp) on a metal dish in a small heap and set alight. The incense must be fully dry, otherwise it will smoulder and not burn.

Lavender Incense Sticks

This incense is made from the flower-head stalks once the buds have been dried and removed.

dried lavender stems	1 cup warm water
1 tbsp (1⅓ US tbsp) saltpetre	

Dissolve the saltpetre in the warm water and soak the lavender stems in it for 30 minutes. Dry out and light for a slow smouldering scent.

Pot-Pourri Cushions

Sew together some muslin to make a bag, fill with your favourite pot-pourri blend, and then cover it with material to match your lounge or chair. Whenever anyone leans back against it the fragrance will be released. Why not make a cushion for each chair with different blends and occasionally swap them around?

Herbs in the Kitchen

Sweet scents in the kitchen are a continual delight. What better to combat cooking smells, freshen up a broom cupboard, duster drawer and odds-and-ends drawer than a perfumed herb bag. Make the bags from squares of coarse material, such as hessian, gathered across the top and tied with a ribbon.

Hang them around the room, over the backs of chairs, and in the cupboards, squeezing them whenever you walk past to release their scent.

Drawer bags can be made from pieces of floral cotton about 15 cm (6″) square. They can be adorned with lace and appliqués and filled with fragrant and insect-repelling herbs.

Try any of the following recipes and prepare as you would a dry pot-pourri.

Lemon Mix

75 g (2²/₃ oz) dried lemon verbena

25 g (1 oz) dried lemon scented geranium leaves

25 g (1 oz) dried peppermint

2 pieces dried lemon peel, powdered

2 tbsp (2²/₃ US tbsp) orrisroot powder

Tangy Green Mix

50 g (1³/₄ oz) dried lemon balm

25 g (1 oz) dried bay leaves, crumbled

15 g (¹/₂ oz) dried spearmint

15 g (¹/₂ oz) dried lovage

10 g (¹/₃ oz) dried pennyroyal

2 tsp freshly ground cloves

2 tbsp (2²/₃ US tbsp) orrisroot powder

Lavender and Lemon Mix

> 50 g (1¾ oz) dried lavender buds
>
> 25 g (1 oz) dried lemon balm
>
> 15 g (½ oz) dried lemon scented geranium leaves
>
> 15 g (½ oz) dried mint scented geranium leaves
>
> 1½ tbsp (2 US tbsp) orrisroot powder

Cupboard Freshener

> 50 g (1¾ oz) dried spearmint leaves
>
> 50 g (1¾ oz) dried pennyroyal
>
> 50 g (1¾ oz) dried tansy
>
> 50 g (1¾ oz) dried lovage
>
> 25 g (1 oz) orrisroot powder
>
> 2 tsp freshly ground cloves

Drawer Freshener

> 50 g (1¾ oz) dried peppermint leaves
>
> 50 g (1¾ oz) dried pennyroyal
>
> 50 g (1¾ oz) dried tansy
>
> 25 g (1 oz) dried lavender buds
>
> 2 tsp freshly ground cloves
>
> 25 g (1 oz) orrisroot powder

Scented Oven

Place a sprig of rosemary or any other savoury herb in the oven whenever you bake; its fragrance will soon fill the kitchen.

Aromatic Oven Gloves

Sew a pocket onto the palm of each glove and fill with rosemary, thyme, lemon balm or crushed spices. Their scent will be released when handling warm utensils.

Fragrant Tea Cosy

Pockets sewn onto the side of your tea cosy can be filled with fragrant herbs whenever you brew a pot of tea. The warmth of the teapot will activate their perfume making your cuppa an aromatic delight. Try herbs such as lemon verbena, jasmine or rose petals.

Eliminating Kitchen and Other Household Odours

Kitchen

- to dispel cooking odours and other kitchen smells, burn a few sprigs of southernwood on a little metal dish
- a cup of rosemary vinegar, or other herbal vinegar, placed close to the stove helps to eliminate cooking odours
- two or three sprigs of spearmint kept in a glass jar of water in the refrigerator eliminates smells
- deodorise your refrigerator after cleaning by placing an open packet of bi-carb soda inside it. This will keep it smelling fresh for up to three months
- rub mustard into hands to remove the smell of fish
- remove onion odour from hands by rubbing with salt
- deodorise jars and bottles by pouring in a solution of water and dry mustard and allow to stand for several hours
- eliminate odours from wooden cutting boards by rubbing them over with salt. Wipe clean with cold water and then rub over again with a handful of fresh spearmint or peppermint to leave them smelling fresh and clean
- deodorise bread boxes or bins by wiping over with 3 tablespoons (4 US tbsp) of bicarb soda dissolved in 1 litre (2 pt) of warm water, or a herb vinegar such as rosemary or thyme.

Household

- deodorise carpets by sprinkling bicarb soda over them. Leave for one or two hours, vacuum off and sprinkle a few drops of your favourite aromatic oil around, or add it to the vacuum cleaner bag
- a few slices of lemon placed in a shallow dish of water will remove the smell of cigarette smoke from a room
- all the pot-pourri and essential oil recipes will remove the staleness and other unpleasant odours from a room
- keep drawers fragrant fresh by lining them with sheets of aromatic homemade paper (see chapter 1, Homemade Paper)
- prevent musty drawers by placing a herb sachet in them. They are smaller versions of the kitchen herb bags and can be made from printed cotton or organdie.

Fresh Drawer Mix

The following recipes will keep drawers fragrant fresh.

25 g (1 oz) lavender flowers	1 tsp freshly ground nutmeg
25 g (1 oz) hyssop	1 tbsp (1¹/₃ US tbsp) orrisroot
15 g (¹/₂ oz) lemon balm	powder

Prepare as you would a dry pot-pourri.

Rose Mix

4 tbsp (5¹/₃ US tbsp) dried red rose petals	8 drops rose oil
	¹/₂ tsp orrisroot powder

Thoroughly mix all the ingredients in a ceramic bowl, fill your sachets and use immediately.

- eliminate shoe odour by sprinkling powdered herbs in them each evening. Use dried chamomile, hyssop or pennyroyal, and reduce to a powder by rubbing through a fine sieve.

Bedroom Herbs

Fragrant herbs will sweeten your clothes, help you to sleep, and lift the spirits of someone who is ill.

Sweet Bags

Like the herb sachets, these are small versions of the kitchen bags and are usually dainty and made from attractively printed material. Place in drawers amongst lingerie, jumpers and linen, or sew on a ribbon loop to attach to hangers under dresses and shirts.

Choose a more robust material, such as muslin, for sachets to slip in shoes, boots and suitcases.

Lavender Mix

4 tbsp (5¹/₄ US tbsp) dried lavender flowers	3 drops lavender oil
¹/₂ tsp orrisroot powder	

Lemon Mix

4 tbsp (5¹/₄ US tbsp) dried lemon verbena	6 drops lemon oil
¹/₂ tsp orrisroot powder	

Fragrant Mint

> 4 tbsp (5¼ US tbsp) dried pennyroyal
>
> ½ tsp orrisroot powder
>
> 3 drops pennyroyal oil

This recipe may also be made with spearmint or peppermint. Prepare all three blends as you would *Rose Drawer Mix*.

Fragrant Powder Sachets

These are smaller versions of the sweet bags that are filled with finely ground herbs. They can be tacked to the lining of coats or jackets to keep them smelling clean and fresh.

Sweet Ambrosia

> 1 cup dried lavender flowers
>
> 1 cup dried elder flowers
>
> 1 cup dried red rose petals
>
> 1 cup dried rose geranium leaves
>
> 1 tbsp (1⅓ US tbsp) orrisroot powder
>
> 1 tbsp (1⅓ US tbsp) powdered gum benzoin
>
> 1 tbsp (1⅓ US tbsp) sandalwood powder
>
> 2 tsp fine salt
>
> 4 drops bergamot oil
>
> 4 drops patchouli oil

Grind each flower and herb separately, using a pestle and mortar. When they have been reduced to a rough powder, pass each one through a fine sieve.

Put 1 tablespoon (1⅓ US tbsp) each of the powdered herbs in a ceramic bowl, add the rest of the ingredients, mix thoroughly, and pass the whole of the mixture through a fine sieve. Put 2-3 tablespoons (3-4 US tbsp) of powder into each sachet.

Scented Coathangers

When making fabric-covered coathangers, add dried herbs, such as rose and lavender, with the padding.

Lavender and Rose

> 3 tbsp (4 US tbsp) dried lavender flowers
>
> 1 tbsp (1⅓ US tbsp) dried red rose petals
>
> ½ tsp orrisroot powder
>
> 3 drops lavender oil

Minty Citrus Blend
(This mix is suitable for a man's hanger.)

> 2 tbsp (2²/₃ US tbsp) dried
> lemon verbena
>
> 1 tbsp (1¹/₃ US tbsp) dried
> lemon thyme
>
> 1 tbsp (1¹/₃ US tbsp) dried
> spearmint

> ¹/₄ tsp orrisroot powder
>
> 4-6 drops lemon oil

Prepare both recipes as you would the *Fresh Drawer Mix*.

Pomanders
Place them in cupboards and drawers, or tied with a ribbon they can be hung in wardrobes.
To make a pomander you will need:

> 1 large, thin-skinned orange
> 1 jar whole cloves

> 1 tbsp (1¹/₃ US tbsp) freshly
> ground cinnamon
>
> 1 tbsp (1¹/₃ US tbsp) orrisroot
> powder

Gently knead the orange in your hands to soften the skin. Make a ring of holes around it with a wooden skewer, and press a clove into each hole. Continue to do this, working in circles towards each end, until the entire orange skin is covered.

Mix together the cinnamon and orrisroot powder. Roll the orange around in it until as much of the mixture as possible clings to it. Wrap in tissue paper, place in a brown paper bag, and leave to cure in a dry, dark, airy place for 3-5 weeks. During this time the orange will dry out completely and shrink slightly.

Finally, take the orange from the tissue and shake off any surplus powder. Your pomander is now ready for use. Tie a velvet ribbon, in old-fashioned colours of crimson or soft pink, around it to suspend it. The scent should last for many years.

Sleep Pillows

What better way to drift off to sleep each night than to the fragrant aroma of soothing herbs. You can stuff a pillow full of sweet smelling dried hops, or place the herbs in a flat muslin bag to slip inside your pillow case.

A herb bag, 30 x 20 cm (12″ x 8″), requires about 70 g (2½ oz) of dried herbs to fill it; smaller sachets that can be hung from the headboard or put inside the pillow slip are far more practical.

Use only dry, crisp herbs and flower petals. Mix all the ingredients together in a ceramic bowl, then put into an air-tight glass container and store in a dark, dry, warm spot for 6 weeks. Shake occasionally to help cure the mixture.

Hop Pillow

> *250 g (9 oz) hops*
> *50 g (1¾ oz) woodruff*
> *50 g (1¾ oz) agrimony*

> *50 g (1¾ oz) southernwood*
> *50 g (1¾ oz) rosemary*
> *1½ tbsp (2 US tbsp) orrisroot powder*

Country Mix

> *25 g (1 oz) woodruff*
> *15 g (½ oz) agrimony*
> *15 g (½ oz) southernwood*
> *2 tbsp (2⅔ US tbsp) marjoram*

> *2 tbsp (2⅔ US tbsp) rosemary*
> *3 bay leaves, crumbled*
> *1 tbsp (1⅓ US tbsp) orrisroot powder*

Lavender and Rose

> *25 g (1 oz) lavender flowers*
> *25 g (1 oz) red rose petals*
> *15 g (½ oz) lemon balm*
> *2 tbsp (2⅔ US tbsp) chamomile*

> *1 tbsp (1⅓ US tbsp) rosemary*
> *2 tsp freshly ground cloves*
> *1 tbsp (1⅓ US tbsp) orrisroot powder*

FIRST AID & TOILETRIES   5

Note to readers

None of the following first aid preparations are intended to replace any prescribed medicines by your doctor, but are merely 'household simples' that provide a natural alternative for first aid situations.

Further, the author does not directly or indirectly dispense medical advice or prescribe the use of the various ointments without medical approval. The intent is to offer information that you may wish to explore as a natural alternative to chemicals.

Should you use the following information you are prescribing for yourself. This is your right, but the author and publisher assume no responsibility for you doing so.

TOILETRY PREPARATIONS

> Before making any of the following toiletry recipes, refer to the general instructions and advice at the beginning of chapter 4.

HERBAL FIRST AID

Herbs can be used successfully as a first aid measure to treat minor complaints, such as bruises, cuts and scratches, sores, etc. They enable you to deal with everyday situations, exerting their beneficial effect as an antiseptic or ointment preparation.

A herbal ointment is a mixture of oils and herbs (introduced as an infusion or decoction — see chapter 8), stiffened with beeswax and lanolin. This will give you a penetrative ointment.

For a non-penetrative ointment, such as is required in the treatment of muscular pain and chest complaints, substitute vaseline for the beeswax and lanolin.

The following ointments can be used for most first aid situations around the home:

Calendula — Pot Marigold (Penetrative):
Use to treat —

burns, cracked lips, weeping sores, wounds, cuts, abrasions, scratches, grazes, and sore nipples. All general skin irritations, including itching, mosquito bite itch and insect bites, and to ease sprains, strains, wrenched ankles and painful swellings.

Chickweed — (Penetrative):
Use to treat —

haemorrhoids, chapping and cracking of skin, nettle rash, itching and eczema.

Garlic — (Penetrative):
Use to treat —

muscular strain and cramp and bruises.

Garlic — (Non penetrative):
Use to treat —

rheumatism, arthritis and aching and painful joints. It can also be applied for backache and all muscular pain.

Eucalyptus — (Non penetrative):
Use to treat —

nasal and chest congestion due to colds.

Apply to nostrils and rub well into chest when retiring.

Penetrative Ointment

Calendula —

4 tbsp (5¹/₃ US tbsp) dried flower petals

40 ml (1¹/₃ fl oz) distilled water

5 g (¹/₆ oz) beeswax

10 g (¹/₃ oz) anhydrous lanolin

150 ml (5 fl oz) almond oil

3 drops tincture of benzoin

(friar's balsam from the chemist)

Add the measured amount of the dried herbs and almond oil (or other cold-pressed seed oil) to a saucepan, bring to the boil, and then reduce to a simmer for 10-15 minutes. Remove from heat, strain and add required quantity to recipe.

In a double saucepan, melt the beeswax and lanolin over a medium heat until completely liquid. Warm the herbal oil and add 75 ml (2½ fl oz), plus the distilled water, to the waxes. Stir until well blended, remove from heat and add tincture of benzoin. Beat with a wooden spoon or electric mixer until cool and of a creamy texture. Store in tightly capped, sterilised glass jars.

Chickweed

60 g (2 oz) fresh chickweed	*10 g (¹/₃ oz) anhydrous lanolin*
20 ml (²/₃ fl oz) distilled water	*150 ml (5 fl oz) almond oil*
20 ml (²/₃ fl oz) aloe vera juice	*3 drops tincture of benzoin*
5 g (¹/₆ oz) beeswax	*(friar's balsam)*

Prepare as for calendula ointment, adding the aloe vera with the distilled water.

Garlic

2 cloves garlic	*10 g (¹/₃ oz) anhydrous lanolin*
40 ml (1¹/₃ fl oz) distilled water	*75 ml (2½ fl oz) almond oil*
	3 drops tincture of benzoin
5 g (¹/₆ oz) beeswax	

Crush the garlic cloves and seal in an air-tight jar with the oil for 7 days. Strain and add oil to the recipe, preparing it as for calendula ointment.

Non-Penetrative Ointment

Garlic —

2 cloves garlic	*100 g (3½ oz) vaseline*

Crush cloves and mix well with vaseline. Store in a jar that has a tight fitting lid. Leave for a week before using.

Rub well into affected area.

Eucalyptus

5 ml (1 tsp) eucalyptus oil	*100 g (3½ oz) vaseline*

Put the vaseline in a small bowl over boiling water and stir until melted. Remove from heat, add eucalyptus oil and stir until well blended. Pour into a sterilised glass jar and allow to cool.

Herbal Antiseptic

A few drops of lavender or tea tree oil in a cup of water makes an all purpose antiseptic for cuts, scratches, etc.

Alternatively, blend equal parts of eucalyptus oil and aloe vera juice and store in an amber bottle until needed. Best used for grazes and scratches.

Apply either antiseptic with clean lint or cotton wool.

TOILETRIES

BATHROOM PRODUCTS

Herbs can be used in many ways in bathroom products, and will exert their beneficial properties upon the skin.

Toilet and Herb Soaps

Recipes for toilet and herb soaps, bubble bath liquid, bath washing liquid and shower washing gel can be found in Chapter 4 (In The Bathroom).

Bath Oil

Essential herb and flower oils can be added singly, or in combination, and blended with a carrier oil, to your bath.

Blend 30 drops of your favourite oil with 40 ml (1⅓ fl oz) of almond oil and store in an amber glass bottle in a cool spot. Add ten drops of the oil while the taps are running.

Try the following combinations:

Calming and Relaxing

2 drops rosemary	*2 drops bay leaf*
4 drops lime flower	*2 drops pennyroyal*

Ease Tired and Aching Muscles

5 drops hyssop	*2 drops bay leaf*
3 drops rosemary	

Bath Vinegar

Herbal vinegar added to the bath acts as an astringent, and will refresh and soften the skin.

Choose your favourite herbs, and make the vinegar according to the directions in chapter 8 (Herbal Vinegars).

Since it will keep indefinitely, you can adjust the recipe and make a larger quantity, storing it in a tightly sealed glass bottle ready for use.

Shampoo

Homemade shampoos contain no synthetics or chemicals, are completely biodegradable, and can be formulated to suit your hair type.

The herbs are first prepared as an infusion, then added to a base recipe to make the shampoo.

Prepare the 'herbal infusion' as directed in chapter 8, substituting the ingredients for 3 tablespoons (4 US tbsp) of dried herb and 1½ litres (3 pt) of boiling water.

Base Recipe

> *100 g (3½ oz) basic soft soap,*   *juice of 1 lemon*
> *grated (see chapter 3)*   *1½ l (3 pt) herbal infusion*

Add the grated soap, lemon juice and 350 ml (12 fl oz) of the herbal infusion to a saucepan, stir, and bring gently to the boil. Reduce heat and continue to stir until the soap has completely dissolved. Add the remaining herbal infusion, stirring until well blended. Bottle for future use.

Shampoo Herbs

Fair hair	*chamomile* — has a lightening effect and is healing to scalp irritations
Dark hair	*rosemary or sage*
Dandruff	*nettle, parsley, peppermint, rosemary and thyme. Thyme, rosemary and peppermint* combined not only control dandruff but act as a scalp and hair tonic
Oily hair	*sage, yarrow, rosemary and lime flowers*
Healing	*chamomile, parsley, rosemary and peppermint*

Conditioning Rinses

Choose herbs singly, or in combination, from the shampoo herbs and prepare an infusion as directed in Herbal Infusion, chapter 8, adjusting the quantity to 2 litres (4 pt).

After washing your hair, rinse thoroughly with clean water and pour over the herbal rinse. Repeat several times, each time massaging well into your hair.

After-Bath Body Powder

An aromatic body powder that is a delight to use after bath or shower.

90 g (3 oz) French chalk

50 g (1¾ oz) corn flour

4 g (2¼ drams) magnesium carbonate

6 g (3½ drams) calcium carbonate

1 tbsp (1⅓ US tbsp) orrisroot powder

3 tbsp (4 US tbsp) lovage or rosemary water

1 tsp rose oil

Mix the dry ingredients. Add the essential oil and herb water (see Herbal Infusion, chapter 8) and mix them thoroughly until the powder feels dry. Extra oil can be added if the scent is not strong enough, but take care not to get the mixture too wet. If it does become a little too wet adjust by adding more powder, a little at a time. Once dry, sieve twice and store in a plastic bottle with holes punched in the lid.

Herbal Deodorant

An effective deodorant can be made by steeping herbs in cider vinegar. It will have both a subdued perfume and antiseptic properties, and will keep you feeling fresh and odour-free.

Liquid Deodorant

5 tbsp (6½ US tbsp) fresh rosemary

cider vinegar

4 tbsp (5⅓ US tbsp) fresh thyme

Prepare the deodorant as directed for *Herbal Vinegar* in chapter 8, then dilute every 5 ml (1 tsp) of vinegar with 40 ml (1⅓ fl oz) of distilled water.

After washing and drying under arms, dab on the deodorant and allow to dry.

Other herbs suitable for making deodorant vinegars are: lavender, sage, lovage, eau-de-cologne mint, spearmint, scented geranium leaves, marjoram and honeysuckle.

Cleaning Your Teeth

A quick and simple tooth cleaner can be made by combining sea salt and bicarbonate of soda.

2 tbsp (2²/₃ US tbsp) fine sea salt

3 tbsp (4 US tbsp) bicarbonate of soda

Mix the two ingredients together and store in an air-tight jar. To use, shake a little into the hand and pick it up on your brush.

Herbal Toothpowder

15 g (½ oz) fresh sage leaves

25 g (1 oz) coarse sea salt

10 g (⅓ oz) fresh peppermint leaves

Mix the herbs and salt together and then spread them out on a baking tray. Place in a preheated oven (150°C/300°F) for 20 minutes, or until the herbs are crisp and dry. Reduce the mixture to a powder by rubbing through a fine metal sieve. Store in an air-tight jar.

Use instead of toothpaste, with a soft bristle brush.

Protective Hand Cream

Use this barrier cream to protect your hands from drying detergents and grime. Smooth it into your hands before beginning rough work or immersing them in washing water.

25 g (1 oz) anhydrous lanolin

10 g (⅓ oz) beeswax

75 ml (2½ fl oz) almond oil

40 ml (1⅓ fl oz) lemongrass infusion

2 tsp lemon juice

3 drops tincture of benzoin (friar's balsam)

Prepare the lemongrass infusion as directed in Herbal Infusion, chapter 8.

Melt the wax and lanolin in a double boiler over a low heat. When completely liquid, stir in warmed almond oil and lemongrass infusion until well blended. Remove from heat, add lemon juice and tincture of benzoin and beat continually until cool and of a creamy texture. Store in a sterilised, screw-top, glass jar.

Massage thoroughly into hands whenever required.

Herbal Make-up

It is impossible to make the same vast range of colour cosmetics at home that is available commercially. However, it is possible to make a few preparations that will adequately suite your basic needs.

Lip Gloss

> 1 tsp alkanet root
>
> 60 ml (2 fl oz) almond oil

> 15 g (½ oz) beeswax
>
> 3 drops tincture of benzoin

Crush the alkanet root and place it in a glass jar. Add warmed almond oil, seal tightly, and place where it will receive plenty of sunlight for 14 days. Strain the oil and set aside.

Melt the beeswax in a double pan over a gentle heat. When completely liquid stir in the warmed alkanet oil, remove from heat, add tincture of benzoin and beat continually until cool and creamy. Store in a sterilised glass jar.

Use as a gloss for the lips.

Lipstick

> 2 tbsp (2⅔ US tbsp) alkanet root
>
> 120 ml (4 fl oz) sesame seed oil

> 30 g (1 oz) beeswax
>
> 3 drops tincture of benzoin

Prepare as for lip gloss, and when cool pour into sterilised glass jars or empty lipstick containers.

Use to colour lips.

Eye Shadow

> 120 ml (4 fl oz) almond oil
>
> 30 g (1 oz) selected herb or flower for colouring

> 30 g (1 oz) beeswax
>
> 3 drops tincture of benzoin

Gently simmer selected herb or flower in the almond oil until it's dye has been released. Remove from heat and when cool strain through muslin cloth.

Melt wax in a double pan over a low heat. When liquid, add herb oil, mixing thoroughly, then remove from heat, add tincture of benzoin, and beat continually until cool. Store in a sterilised glass jar.

To colour your eye shadow use any of the herbs listed in the

Herbal Dye Chart, chapter 4 (Natural Dyes). Other plants to use are beetroot, blackcurrant, and black or blue malva flowers. Alternatively, you may wish to experiment with different plants to see what colours they will yield.

Male Toiletries

Shaving

For wet shaving, use either the shaving cream or soap in chapter 3, after which the skin should be toned with a herbal aftershave lotion.

Herbal Aftershave Lotion

1½ tbsp (2 US tbsp) chopped sage leaves

1½ tbsp (2 US tbsp) rosemary leaves

1½ cups cider vinegar

1½ cups witch hazel (from the chemist)

Prepare the herbs and cider vinegar according to the directions in chapter 8 for *Herbal Vinegar*, then stir in the witch hazel. Store in a tightly capped glass bottle.

Pat onto the skin with a piece of cotton wool.

Aftershave Cologne

A suitable cologne can be made by steeping a sharp-scented herb in vodka. Suitable herbs to choose from are:

angelica, bergamot, fennel leaves, geranium leaves (lime, lemon, lemon-rose, mint-rose, spicy, etc), juniper, lemon balm, lemon verbena, lime flowers, peppermint, pine needles or thyme.

Two-thirds fill a wide-mouthed glass jar with vodka. Add as many fresh herbs as you can until no more can be forced into the jar and tightly seal. Allow the mixture to stand for about six weeks, or until the aromatic essence has left the herbs. Strain, drip through filter paper and store in a tightly sealed glass bottle.

Beards and Moustaches

Beards and moustaches also require attention and regular grooming. Wash them every time you wash your face, using the finger tips to massage the skin underneath. Dry, then apply a herbal oil such as basil, rosemary or sage.

To apply the oil, rub a little of it between the palms of the hand and stroke a good quality hairbrush over them to pick up the oil. Brush through the beard and moustache — if the latter is too difficult use a toothbrush.

Aftershave Moisturiser
If facial skin feels dry after shaving and toning, lightly dab apricot kernel oil onto dry skin.

OUTDOORS

Sunscreen Cream

This is the cream my family uses when spending a day in the sun and it has so far proved successful in blocking about half of the sun's burning rays. However, like us, you must also take a sensible approach towards dress when using this cream, and it may then prove beneficial to you.

¾ cup very strong tea	*20 ml (²/₃ fl oz) carrot oil*
¼ cup anhydrous lanolin	*¼ cup calamine powder*
¼ cup sesame oil (cold pressed)	*8 drops rosemary oil*

In a ceramic teapot brew ¾ cup of very strong tea, using 3 teabags and infusing for 30 minutes.

Put the lanolin, sesame oil, carrot oil, calamine powder and ¼ cup of tea in an electric blender and whirl at low speed. When completely blended increase the speed and pour in the remaining tea in a thin, steady stream. Add the rosemary oil last. Store in a suitable glass jar with a tight-fitting lid.

Apply to exposed areas of skin whenever out of doors for prolonged periods of time. However, you must still be mindful of the severe and irrepairable damage that can be caused through excessive exposure to the sun's rays.

Sunburn Lotion
Ideal for mild sunburn and preventing further moisture loss.

50 g (1¾ oz) glycerine	*10 ml (¹/₃ fl oz) wheatgerm oil*
20 ml (²/₃ fl oz) aloe vera juice	*10 ml (¹/₃ fl oz) jojoba oil*
20 ml (²/₃ fl oz) rose water	*5 drops tincture of benzoin*

Mix together all ingredients, then beat vigorously until completely emulsified. Store in a tightly sealed amber glass bottle.

Shake before use and apply generously, but gently, over affected areas of skin.

Insect Repellent

This repellent is disliked by both mosquitoes and flies when rubbed onto the skin.

> *125 ml (4 fl oz) elder flower*   *5 ml (1 tsp) vodka*
> *infusion*
>
> *6 drops lavender oil*

Prepare the elder flower infusion as directed in Herbal Infusion, Chapter 8.

Gently warm the elder flower infusion and set aside. Dissolve the lavender oil in the vodka and then mix thoroughly with the warm infusion.

Apply generously to exposed skin.

Other effective repellents

See Mosquitoes, chapter 6, for effective body repellents and other control methods.

Alternatively, fresh lemon balm, pennyroyal or basil leaves when bruised and rubbed onto the skin will keep mosquitoes and flies at bay, as well as relieving mosquito bite itch.

Insect Bite Soothers

Fresh hyssop, plaintain, rosemary, lemon balm, pennyroyal, basil or yarrow will relieve insect bites and stings when bruised and rubbed into the affected area. A cornflower infusion applied generously also takes the itch out of insect bites.

Calendula ointment (see Herbal First Aid, this chapter) is also excellent for relieving all general skin irritations, including itching and mosquito and other insect bite itch.

6 FAMILY PETS & HOUSEHOLD PESTS

PET CARE WITH HERBS

Before the human race domesticated animals for pets they would seek out their own remedies and eat those plants that made them healthy. Now that they live in our environment, those plants which help to keep them well are not always readily available.

You can add these herbs as supplements to their diet as well as using them to treat simple ailments. If your pet is given herbs in its diet at an early age you will find it will enjoy them.

Herbs such as garlic, parsley, watercress and dandelions can be added finely chopped or minced, and combined with some grated raw carrot. To this you can add wheatgerm flakes, yeast, cod-liver oil and kelp. Mix ingredients thoroughly with raw meat.

This diet is only a guideline and can be varied from day to day, by just including some or all of the ingredients.

Grooming and Flea Control

Wash the dog regularly with a soap which will help to control fleas (see *Dog Washing Soap*, chapter 3) and then rinse with a herbal flea wash, working it well into the coat. Pay particular attention to the areas around the ears, neck, tail, backbone and under the legs, then comb through a solution of equal parts pennyroyal oil (see chapter 8) and eucalyptus oil. A mixture of garlic and elder lotion is also effective, as is 1 teaspoon of cajeput oil to every 5 tablespoons (6½ US tbsp) of warm water. Both the oils and the lotion can be used on cats too — apply to their fur and rub through with your fingers.

In between washes dust your pet's hair with a herbal flea powder, leave it for half an hour, then brush or comb it out with a fine-tooth comb. Sweep up all dead and stunned fleas and burn them. Brushing will also remove loose hair, grass seeds, burrs, etc, and should be done frequently with long-haired pets.

Pet Hygiene

All sleeping areas and bedding should be kept dry, frequently washed or dry-cleaned, and aired at least once a week in the sunlight. Dried pennyroyal or pennyroyal powder scattered or sprinkled on pet bedding will repel fleas, as will a bedding of cedar wood shavings or pine needles. The bedding material can be stuffed into a large cushion with the dried insect-repellent herbs. Include pennyroyal, stinking roger, wormwood, tansy, native peppermint or rue; the latter should be excluded from cat's bedding, as they usually hate it. Replace the rue with catmint and any of the other herbs, and your cat will be very pleased. Lavender and tansy oil spinkled on bedding will also help to repel fleas.

Dust carpets and mats with flea powder, leave for an hour and vacuum; dispose of any dead or stupified fleas by burning them. Cajeput oil, diluted with warm water (see Grooming and Flea Control) can be sprayed on carpets, lounge chairs, etc, to rid your home of fleas.

Regularly empty and clean water bowls to eradicate mosquito larvae, and frequently wash food dishes to prevent a crusty build-up of food particles that attracts flies. Remove any faeces and keep the area in which pets are confined clean and dry; it's the smell and moisture that will bring the flies around. Bury dog faeces, don't compost them; animals are prone to too many diseases.

Flea Repellents

Garlic and Elder Lotion

This will not only help in the control of fleas, but is a very effective lotion for the treatment of mange.

> *100 g (3½ oz) garlic*
> *400 g (14¼ oz) elder leaves*
> *40 ml (1⅓ fl oz) olive oil*
>
> *40 ml (1⅓ fl oz) methylated spirits*
> *water sufficient to make 2 litres (4 pt)*

Chop garlic, place in a glass jar, add olive oil, seal with an airtight lid and steep for 48 hours.

Place elder leaves in an enamel saucepan, add sufficient water to cover, bring to the boil, then reduce to a simmer for 45 minutes with the lid on the pan. Remove from heat and steep overnight, then strain through muslin cloth, squeezing all liquid from the herbs. Stir in garlic oil, methylated spirits and sufficient extra water to make up 2 litres (4 pt), then strain. Store in well sealed bottles in a dark, cool cupboard for up to 3 months.

Rub thoroughly through your pet's fur after washing.

Garlic and Elder Flea Wash

A combination pet wash and flea control lotion that will also aid in the treatment of mange.

> 40 ml (1⅓ fl oz) garlic oil
> 1 l (2 pt) elder leaf infusion
>
> 15 g (½ oz) basic soft soap, grated (see chapter 3)
> 1 l (2 pt) boiling water

Prepare the garlic oil and elder leaf infusion as for *Garlic and Elder Lotion*.

Dissolve the grated soap in the boiling water, then add the garlic oil and elder leaf infusion, stirring until well blended. Store in tightly sealed bottles in a dark, cool cupboard for up to 3 months. Keep out of the reach of children.

This mixture may be further diluted in the ratio of 1 litre (2 pt) of water to every 2 litres (4 pt) of flea wash.

Herbal Flea Wash

This should be used to thoroughly rinse your pet's fur after washing with a flea control soap.

> ½ cup fresh pennyroyal
> ½ cup fresh southernwood
>
> 1 l (2 pt) water

Place the herbs in an enamel saucepan, add the water, bring to the boil, remove from heat, cover, and infuse overnight, then strain through muslin cloth, squeezing all liquid from the herbs. Use within 24 hours, unless kept in the refrigerator; it will then last for up to 7 days.

Liquid Washing Soap

Use in place of the dog washing soap and herbal flea wash.

> 1 l (2 pt) herbal flea wash
>
> 15 g (½ oz) soft soap, grated (see chapter 3)

Bring the herbal flea wash to the boil, remove from heat and stir in the soft soap until it has completely dissolved. Allow to cool before use.

A stronger wash can be made by replacing the southernwood with pyrethrum flowers. If you are unable to get pyrethrum flowers, use twice as many feverfew flowers instead. However, you must never boil pyrethrum flowers as the fumes are toxic; infuse in hot water instead.

Pet Bedding Spray

To eliminate the possibility of fleas reinfesting pets, when airing their bedding, spray immediately with the following insecticide as soon as it's put outside in the sun.

> 5 tbsp (6½ tbsp) feverfew
> flowers
>
> 1 l (2 pt) boiling water
>
> 20 ml (²/₃ fl oz) methylated
> spirits

Place the flowers in a ceramic bowl, add the boiling water, cover, steep overnight, then strain through muslin cloth, squeezing all liquid from the herbs. Pour into a pump-spray bottle, add the methylated spirits and shake vigorously so that it is well blended.

Herbal Flea Powder

Use this for flea control on pets and for dusting bedding, carpets, mats and chairs.

> 100 g (3½ oz) dried
> pennyroyal
>
> 100 g (3½ oz) powdered
> fennel seed
>
> 100 g (3½ oz) dried
> pyrethrum flowers

Reduce dried fennel seeds to a powder with a pestle and mortar, and powder the other herbs by rubbing them through a fine wire sieve. Thoroughly mix all ingredients and store in a plastic bottle with holes punched in the lid.

Pyrethrum flowers can be replaced by feverfew flowers, using 200 g (7 oz) of the dried herb.

Herbal Pet Cures

The following remedies are only for the treatment of simple complaints and common problems. Any serious complaint should be referred immediately to your veterinarian.

Recipes for ointments and lotions appear at the end of the

text, and all herbal infusions are formulated on the following proportions: 2 tbsp (2²/₃ US tbsp) dried herb to every 300 ml (10 fl oz) boiling water.

Bad Breath	wash gums and teeth with a rosemary infusion using a toothbrush. Also administer a cupful of the infusion each day.
Baldness	loss of hair not caused through disease may be treated by adding dandelion leaves to your pet's daily meal. Also rub the affected spots with 5 drops of eucalyptus oil to every 30 ml (1 fl oz) of castor oil. If the complaint persists, see your vet.
Cuts and Abrasions	bathe with a herbal antiseptic made by infusing equal parts of elder flowers, elder leaves and rosemary. Apply aloe vera ointment after cleansing.
Coughs	minor, persistent coughs can be treated by giving your dog or cat an infusion of equal parts elder flower, thyme and horehound. Administer one cup of infusion mixed with a little honey 2-3 times daily.
Ears	remove foreign bodies from ears by carefully inserting one teaspoon of rosemary oil (see Herb and Flower Oils, chapter 8). Later, dry out the ear gently with swabs and diluted witch hazel (available from your chemist). Canker, which is common in long-eared dogs, can be treated by cleansing the affected ear daily with 3 parts rosemary infusion to one part witch hazel.
Fleas	see Grooming and Flea control.
Flies	odour is the usual cause of flies being

attracted to pets. Check its state of health: diet, eczema, over-active scent glands, infection and sores from scratching fleas, etc.

Mix one teaspoon of eucalyptus oil to every 300 ml (10 fl oz) of warm water and comb this through the animal's fur. Serious complaints or those from no obvious cause should be referred to the vet.

When treating wounds, a little eucalyptus oil brushed around the area will repel flies. Be sure that it doesn't come in contact with the wound.

Lice

wash the animal with dog soap or liquid washing soap, then thoroughly rub and comb in derris-eucalyptus lotion.

Loss of Appetite

usually, the best treatment for an animal that won't eat is not to feed it for two to three days, except for a little grated raw apple and a cup of peppermint infusion twice a day.

Mange

good food rich in vitamins A and B for prevention. Treat by clipping fur away from the affected area and then bathing with warm soapy water to which has been added a few drops of olive oil. Then apply dock balm or garlic oil lotion.

Ticks

although tick paralysis can occur throughout the year, it is most common in late winter and spring.

Search pet's fur daily in areas that are prone to tick infestation, paying particular attention to the head, around the neck, front legs, including between the toes, the nostrils and in and behind the ears. And remember, both the head and body of the tick must come out — use a needle if necessary. Administer an infusion of bracken shoots as an antidote to tick poisoning — it will cure and can save an animal's life if the ticks are found in time. If paralysis symptoms occur, immediately seek veterinary attention.

Apply a tick repellant if you live in a tick-infested area.

Vomiting	administer 40 ml (1⅓ fl oz) of herbal tea, made from equal parts peppermint, rosemary and thyme, three times daily.
Worms	usually worms are the result of a bad diet; they love fats, sugars, eggs and milk. A well balanced diet should prevent them from becoming a problem. A natural preventative is to include one dessertspoon of any of the following foods in your pet's diet each day: grated raw carrot, ground raw pumpkin seeds, finely chopped garlic, melon or grape.

Ointments and Lotions

Aloe Vera Ointment (for cuts, abrasions and wounds)

1 handful dried marigold (calendula) petals
100 ml (3⅓ fl oz) olive oil
60 ml (2 fl oz) aloe vera juice

5 g (⅙ oz) beeswax
10 g (⅓ oz) anhydrous lanolin
3 drops tincture benzoin

Place the calendula petals and olive oil in an enamel saucepan, bring to the boil, reduce the heat and simmer until the herbs are crisp. Strain through muslin cloth, squeezing all oil from the petals.

In a double boiler melt the beeswax and lanolin together over a medium heat. When completely liquid, stir in the aloe vera juice and 90 ml (3 fl oz) of the calendula oil until thoroughly blended. Remove from heat, add tincture of benzoin, pour into a ceramic bowl and beat until a smooth, creamy texture. Store in a tightly capped glass jar. Apply as required.

Dock Balm (treatment of mange)

125 g (4½ oz) chopped dock roots

125 ml (4 fl oz) white vinegar

150 ml (5 fl oz) olive oil

5 g (⅙ oz) beeswax

10 g (⅓ oz) anhydrous lanolin

3-4 drops tincture benzoin

40 ml (1⅓ fl oz) distilled water

Place the chopped dock root in an enamel pan, add the vinegar, bring to the boil and cook until vinegar has nearly boiled away, then strain. Place the dock back in the pan, add the oil, bring to the boil and then reduce to a simmer for 30 minutes, then strain.

In a double boiler melt the beeswax and lanolin over a medium heat, stir in the distilled water and 75 ml (2½ fl oz) of herbal oil until thoroughly blended. Remove from heat, add tincture of benzoin, pour into a ceramic bowl and beat until a smooth, creamy texture. Store in a screw-top glass jar.

Derris-eucalytpus Lotion (treatment of lice)
Powdered derris is an organic substance obtained from the pulverised root of the parent plant. It is usually sold as a powder available from any garden suppliers or the garden section of most supermarkets. It will break down under sunlight in a few days, but if allowed to enter waterways is deadly to fish; don't pour it down drains and keep it away from fish ponds and dams.

250 g (9 oz) powdered derris

125 ml (4 fl oz) eucalyptus oil

375 ml (12½ fl oz) methylated spirits

4 tbsp (5⅓ US tbsp) dried wormwood

2 l (4¼ pt) boiling water

Combine the derris, eucalytpus oil and methylated spirits in a glass jar, stirring until completely dissolved and liquid. Seal the jar and allow to steep for 5 days, shaking vigorously twice a day.

After the fourth day, place the dried wormwood in a ceramic bowl and add the boiling water. Cover, steep for 12 hours, strain through muslin cloth, squeezing all liquid from the herbs, and blend with the derris solution.

Thoroughly rub the lotion through the animal's fur, making sure that it is wiped onto the skin. If the pet has long hair, cut this short and wipe its skin with the lotion. Repeat 2 weeks later if the symptoms return.

Derris Dusting Powder

100 g (3½ oz) powdered derris	*200 g (7 oz) plain white flour*
200 g (7 oz) powdered feverfew flowers	

Reduce dried feverfew flowers to a powder by rubbing through a fine wire sieve. Combine with the rest of the ingredients and store in a plastic bottle with holes punched in the lid.

For protection against lice, dust the animal's fur and then work well into the skin with your fingers. Repeat every two weeks if necessary.

Garlic Oil Lotion (treatment of mange and flea control)

100 g (3½ oz) chopped garlic	*200 ml (6½ fl oz) olive oil*
100 g (3½ oz) chopped dock root	*40 ml (1⅓ fl oz) methylated spirits*
400 g (14¼ oz) elder leaves	*water sufficient to make 2 litres (4¼ pt)*

Place garlic in a glass jar, add 40 ml (1⅓ fl oz) of oil, seal with an air-tight lid and steep for 48 hours.

Combine dock root and 160 ml (5⅓ fl oz) of oil in an enamel saucepan, bring to the boil then reduce to a simmer for 45 minutes with the lid on the pan. Remove from heat and allow to steep until cool, then strain.

Add elder leaves to an enamel saucepan plus sufficient water to cover, bring to the boil then reduce to a simmer for 45 minutes. Remove from heat, cover, steep for 12 hours, then strain through muslin cloth, squeezing all liquid from herbs.

Combine the herbal infusion, garlic oil, 40 ml (1⅓ fl oz) of dock oil, methylated spirits and sufficient extra water to make up 2 litres (4¼ pt), then strain.

Store in tightly sealed bottles in a dark, cool cupboard and use within 3 months.

Tick Repellent

30 g (1 oz) powdered derris

45 ml (1½ fl oz) eucalyptus oil

80 ml (2⅔ fl oz) pennyroyal-thyme oil (see below)

300 ml (10 fl oz) methylated spirits

2 l (4¼ pt) water

Dissolve the derris in water and the oils in the methylated spirits. Blend all ingredients in a large bottle, shaking vigorously until thoroughly mixed.

Before use, wash your pet first with warm soapy water, or one of the pet washes, then rub repellent thoroughly through the animal's fur, paying particular attention to the neck, head and ears, and the inside of the legs.

Pennyroyal-Thyme Oil

Add one handful of fresh pennyroyal and one handful of fresh thyme (or 2 tablespoons/2⅔ US tbsp each of dried herb) to an enamel pan, cover with olive oil, bring to the boil and reduce to a gentle simmer for 10 minutes or until the fresh herb is crisp. Add required amount to the recipe.

Tick Repellent Powder

100 g (3½ oz) powdered derris

100 g (3½ oz) dried pennyroyal, powdered

Reduce the dried pennyroyal to a powder by rubbing through a fine wire sieve, then mix with the derris. Keep in a plastic bottle with holes punched in the lid.

Dust the powder over the animal's fur, working well in with your fingers.

To be effective, both the liquid and powder repellent must be applied each week and re-applied after swimming or rain.

HOUSEHOLD PESTS

In today's society we have been indoctrinated into believing that we need either chemicals or experts for effective pest control. Sadly, for so many of us there seems to be no other alternative, as the sound lore of our grandparents' day has been lost. However, there is no need to reach automatically for the latest wonder chemical or yell for help to eradicate the many insect pests that bother us from time to time. There are safe, natural, organic methods that will help to control the problem without harming the environment or people.

Pests need to be controlled, rather than totally eradicated. Good housekeeping practices will eliminate the need to continually mass murder everything that creeps, scuttles and crawls. Insect screens will keep out virtually everything from cockroaches to mosquitoes, and regular vacuuming of carpets and mats, behind books and furniture, in cupboards and along shelves will not only clean up fleas, silverfish and cockroaches, but will also disturb their feeding and shelter areas. Hidden pests will become more visible and can be controlled mechanically instead of by the use of harmful sprays.

Most natural methods of insect control tend to repel them rather than kill. However, occasionally more specific remedies may be needed, such as baits or sprays. Use only organic substances that break down quickly and harmlessly. Remember though, they will leave no residual effect to keep killing the pests that have invaded your home. They give you only a respite, not a cure; therefore, prevention is the only effective means of pest control.

Prevention

It is far easier to install flyscreens, incorporate birdwire in the roof of your home and to fit rodent wire than to try and remove the problem once it's there. Other ways in which we encourage pests are: raising the temperature of our houses, inadequate sub-floor ventilation, providing food and shelter (especially in kitchens), and feedsheds. You can keep the temperature of your home comfortable without providing a haven for insects, wipe up spills in the kitchen and keep the compost bin well sealed (even the smallest gap will allow a cockroach entry), and keep animal feed sealed in air-tight containers. As well there are a few other basic steps in prevention:

- look for breeding places and eliminate them
- try the use of deterrents
- don't leave food scraps lying around
- check your compost — if it isn't heating up after 3 days there is something wrong. Correct it before it becomes a breeding ground for pests
- eliminate ceiling and floor cracks, and cracks in and around doors. Seal with silicone and then test with a burning candle — if the flame flickers with all the windows and doors closed there must still be a gap somewhere; find it and seal
- draught extruders on doors also help to keep pests out
- consider planting repellent herbs in pots close to doors and windows or planting a pest control garden. Choose herbs such as lavender, feverfew, pennyroyal, grey-silver hourhound, wormwood, tansy, rue, basil, castor oil plants, peppermint, native pennyroyal, southernwood, santolina, fennel, and flowering elder trees

ANTS

Most householders suffer from ant invasion at some time, especially in dry weather. If they really bother you, hunt down the nest and destroy it, but only as a last resort. Try and discourage them from staying in or entering the house.

- grow pots of pennyroyal, rue or tansy near the kitchen door. Place sprigs of the same herbs on shelves and in cupboards, disturbing the leaves occasionally to release more scent
- sage sprinkled in cupboards is also an effective deterrent
- sprinkle a combination of baking soda and black pepper wherever ants gather; it will soon make them vanish
- slices of lemon strewn in their paths is said to deter them if they can't find an easy way around them
- bone meal (available from garden suppliers) sprinkled around the outside walls of the house and throughout the garden will drive ants away
- keep food scraps and crumbs off ant-accessible areas
- leave repellents or baits on ant trails

Destroying the nest

- place baits on nest or pour homemade insecticides down it
- pour boiling water down the nest until all traces of ants have gone
- Finally saturate the next with homemade *Pyrethrum Insecticide*

To be completely effective both these methods may have to be repeated a number of times, especially when using the pyrethrum spray as it breaks down quickly, leaving no residual.

Ant Bait

> 3 tbsp (4 US tbsp) sugar   250 ml (8 fl oz) warm water
>
> 1 tbsp (1⅓ US tbsp)
> bicarb-soda

Dissolve the sugar in the warm water, allow to cool, then stir in the bicarb-soda, mixing well. Store in a bottle in the refrigerator until needed.

Put bottle capfuls on ant nests and trails; cover if the weather is wet. This type of bait works on the insect's body heat. As they feed on it, or on each other, the body warmth makes them literally blow apart.

Borax Bait

> 2 cups sugar   250 ml (8 fl oz) water
>
> 2 tbsp (2⅔ US tbsp) borax

Mix the sugar and borax together, then dissolve in the water. Use as you would the ant bait. For indoor use, such as in the kitchen, pour into small glass jars loosely filled with cotton wool. Punch holes in the lid and locate where ants are gathering.

Pyrethrum Insecticide

> 250 g (9 oz) powdered dry   methylated spirits, sufficient
> pyrethrum flowers
> 75 ml (2½ fl oz) sesame seed
> oil

Rub dried pyrethrum flowers through a fine wire sieve, then combine with the other ingredients in a glass jar, adding just sufficient methylated spirits to make the whole liquid. Steep for a week, shaking the mixture twice daily, then dilute with 3 parts water to 1 part mixture.

Pour down ant nests or spray departing ants after pouring water down the nest.

COCKROACHES

People tend to shudder at the sight of cockroaches: scuttling around the kitchen after the lights go out, crawling over dishes and invading cupboards and drawers. And for good reason; cockroaches are disease carriers. They may be the direct cause of transferring or spreading such diseases as hepatitis, salmonella, diptheria and dysentery.

Prevention is the first and most important consideration in controlling these pests. Otherwise, remedies, no matter how effective, will be useless if the cockroaches keep returning. Install insect screens, flyscreen doors, rubber strips at the bottom of drawers and seal every crack, crevice or gap you can find. Check around windows, along skirting boards, under lino and anywhere else that they may gain entry.

Don't leave food debris laying about, clean up rubbish in attached garages, store rooms, etc, and seal inside compost bins. Make sure all food containers have air-tight seals and never leave dirty dishes in the sink overnight. Clean up any piles of rubbish outside and keep garbage bins well sealed — a strip of plasticine or Blu-tack around the lip will give a good cockroach-proof seal.

Cockroach Control
_____

A 50-50 mixture of sugar and borax sprinkled under the sink, at the backs of cupboards and drawers, or anywhere cockroaches hide will repel them. Pyrethrum powder is also effective: reduce dried flowers to a powder by rubbing through a fine wire sieve.

Tea tree leaves scattered thickly in runways, in and near cupboards, in drawers, and anywhere else cockroaches are likely to gather, is reputed to repel them.

Pyrethrum or feverfew insecticide (see Fleas) sprayed in crevices, under sinks, in cupboards, along the base of doors, or in breeding places will eliminate them. However, the treatment is not permanent and will need repeating while ever cockroaches persist.

Diatomaceous earth, made from the finely ground skeletons of marine creatures called diatoms, can be used effectively if sprinkled where the insects congregate. The microscopic fragments of the powder grind away at the carapaces, finally causing the cockroach to die from dehydration.

Cockroach Bait
This bait works on the same principle as the ant bait. Body warmth from the insect causes them to blow apart.

> *2 cups cold mashed potato*   *1 tbsp (1¹/₃ US tbsp) baking powder*

Combine the ingredients in a mixing bowl then form in balls the size of a marble. Place 2-3 balls where cockroaches gather, renewing until they are left untouched — the unwanted visitors should then be dead. Leave a few balls around for any future generations.

Remember, baits will only be effective if there is no other accessible food around: dirty dishes, scraps, crumbs, etc.

Cockroach Trap

This trap works very effectively and only requires empty jam jars or the like. Again, there must not be any other easily accessible food around.

Smear the inside top third of the jar with petroleum jelly; this stops them from getting out. Pour a little cooking oil in, just enough to cover the bottom, and add a piece of banana or cake.

Place your traps in cupboards, near the refrigerator, by the garbage bin, or wherever the cockroaches gather. In the morning, pour boiling water into your traps to kill them.

FLEAS
The cause of fleas is most commonly from family pets, and it is quite usual for them to jump straight from dog or cat to human. Eradication and control is discussed in this chapter in Pet Hygiene. However, in addition to the herbal pet remedies and treatments, regular, hard vacuuming is a must. Do the whole house in one day and vacuum lounges, cushions, carpets, mats, up and down curtains, along crevices, and beds and their

coverings. Take animal bedding, cushions, mats, etc, outside and air them in the sun for a day, then vacuum before bringing them back inside. Dust regularly with flea powder or spray cajeput oil solution or *Pet Bedding Spray*. Pyrethrum or feverfew insecticide (spray) is also effective.

Pyrethrum or Feverfew (Spray) Insecticide
Prepare as follows, using either pyrethrum or feverfew flowers. This can be used to effectively control both fleas and cockroaches.

	2½ tbsp (3⅓ US tbsp) dried pyrethrum flowers	*1 l (2 pt) boiling water*
or	*5 tbsp (6½ US tbsp) dried feverfew flowers*	*20 ml (⅔ fl oz) methylated spirits*

Stand pyrethrum flowers in hot water for about an hour, then cover and allow to cool. Strain through muslin cloth, squeezing all liquid from the herbs, stir in methylated spirits and store in a pump-spray bottle.

Prepare with feverfew flowers as above or as detailed for the *Pet Bedding Spray*.

If you have a bad infestation, after vacuuming close all windows and doors and spray liberally with the above spray, paying particular attention to crevices, lounges, etc.

FLIES

Many herbs help to deter flies, including basil, elder, lavender, mugwort, pennyroyal, peppermint, rue, tansy, southernwood and wormwood. However, preventive methods should be the first priority: insect screens on doors and windows, airlocks (closed areas between two doors — see chapter 2) in bad fly areas, rubber strips fitted to the bottom of doors to seal gaps, sliding boards to block off chimneys (they can be easily removed when you light a fire), and rubbish bins kept tightly closed. Check for any possible breeding areas, such as slow compost heaps, damp bedding in dog kennels, chicken pens and yards, rubbish piles, and leaking septic pipes and outlets. Encourage your neighbours to check their properties too, and if you back onto a public reserve check this as well and clean up if necessary.

If flies still manage to find their way in you can't beat the old-fashioned fly swatter, especially once you become deft in its use. Or you can try any one of the following repellents:

- grow pots of basil and place them where flies congregate, such as porches, garages, balconies or doorways. Or leave a vase of fresh basil on a nearby shelf
- rub windows and glass doors with a cloth soaked in lavender oil or water which has been used to boil up onions
- mint grown in pots has a similar effect as basil
- pennyroyal oil brushed onto the woodwork surrounding doorways and kitchen benches will help to keep flies away
- eucalyptus candles (see Candlemaking, chapter 4) burning will repel blowflies. Candles made from any of the other fragrant fly repellent herbs are also effective
- fennel grown close to stables, pet kennels and the back door will help to keep flies away

Fly Traps

Fly Paper

The old-fashioned, long sticky ribbons of paper are just as effective a means of fly control today as they were in grandma's day. They are still available, but you can make your own inexpensively at home. You will need:

75 g (2½ oz) castor oil

180 g (6½ oz) powdered resin

strips of stiff yellow or brown paper

sufficient cardboard discs and string for hanging

Cut strips of paper about 50 mm (2″) wide and the same number of cardboard discs 55 mm (2¼″) in diameter.

In an old saucepan, or tin, warm the castor oil over a gentle heat. Add the resin and stir until dissolved and thick. (Don't overheat as resin is inflammable). Remove from heat and see-saw the strips of paper through the mixture until thoroughly coated. While still wet attach the cardboard disc to the bottom and a length of string to the top (the resin will hold them in place) and hang them up to dry.

Those strips not being used immediately can be rolled up, wrapped in foil, and stored until required.

I have found these fly strips to work just as effectively outside the house in areas such as an open verandah. Combined with pots of fly repellent herbs, it definitely keeps the summer fly population at bay.

Dish Trap

A small dish or saucer is suitable; place in an inconspicuous place. The flies eat the bait and it kills them stone dead.

Combine ½ teaspoon black pepper, 1 teaspoon of brown sugar and 1 teaspoon of cream. Alternatively, place a piece of meat in the bottom of a dish and cover it with *Pyrethrum Insecticide*. Keep the latter out of the way of pets and children. Although it has a low toxicity for animals and humans, it may cause illness.

Soft Drink Bottle Trap

Cut the top off an empty plastic drink bottle at the shoulder, turn the top upside down, insert it back into the bottle and tape into position. Drop in a piece of meat, cover with *Pyrethrum Insecticide*, attach a length of string and hang wherever flies gather. An ideal trap to hang by the kitchen door, in attached garages or semi-open entertainment areas.

The *Pyrethrum Insecticide* can be replaced by water with a little cooking oil floating on the top.

Fly Spray

You can use the *Pyrethrum or Feverfew Insecticide*. However, first, do a patch test on the inner arm of each individual in the household to make sure that no one is sensitive to it; if they are, a rash or redness will develop within 24 hours. Don't use it.

Use pyrethrum only if essential when all other methods will not control plague situations, and only when leaving the house for 2 or 3 hours. It is a very effective spray that breaks down quickly, leaving no residual. Feverfew is safe.

Eucalyptus Fly Spray

I very seldom use a fly spray; however, on the odd occasion when the need has arisen I have had success with the following recipe.

20 ml (²/₃ fl oz) eucalyptus oil   94 ml (3 fl oz) vodka

>   6 ml (1 tsp) bergamot oil   1370 ml (45½ fl oz) distilled
>   10 ml (2 tsp) white vinegar   water

Dissolve the oils with the alcohol, then combine with the vinegar and distilled water, mixing thoroughly. Store in a pump-spray and use as required.

LICE AND NITS

Human lice are passed from one person to another, usually by direct contact. Sharing of combs, brushes, hats, pillows, bed linen, etc, will also spread body and head lice. The only other lice that attaches itself to humans is pubic lice, and is spread through sexual intercourse.

Lice puncture the skin and suck blood, laying eggs — called nits — which attach to the hair. The most common problem is head lice and large outbreaks generally occur amongst school children, since it is in overcrowded conditions that they spread more easily and quickly.

Itching of the head is the first sign that someone may have lice and nits. Thoroughly check the hair, looking close to the scalp — nits are grey-coloured eggs attached to the hair and are just visible with the naked eye. However, they are clearly visible with a magnifying glass.

Treatment
_____

Thyme is a powerful antiseptic which contains thymol and will effectively help in controlling nits. Comb either pure essential oil of thyme or tincture of thyme through the hair with a fine toothed comb.

Oil of thyme is generally available from specialist herb shops and health food shops. If you have difficulty in obtaining it, you can make your own essential oil by distillation or enfleurage (see chapter 8) or a tincture by steeping fresh thyme in alcohol.

To make a tincture, two-thirds fill a wide-mouthed glass jar with vodka and add fresh thyme until no more can be forced into the jar. Seal tightly and leave where it will receive plenty of hot sunlight for three weeks. Strain through muslin cloth, squeezing all liquid from the herbs, then drip through filter paper. Store in a tightly capped bottle.

Use either the oil or tincture daily until the problem clears up.

An old-fashioned treatment is to use a hair rinse made from quassia chips. You will need:

15 g (½ oz) quasia chips   *cider vinegar*

2 l (4¼ pt) of water

Boil the chips for two hours, strain, and add I tablespoon (1⅓ US tbsp) of cider vinegar for every 300 ml (10 fl oz) of liquid. Apply at two-week intervals, three times.

MOSQUITOES

Only the female mosquito sucks blood, and will attack any warm-blooded animal. Regardless, they are still a nuisance, and nothing is worse when you are just dozing off at night than to hear that familiar buzz. In years gone by, I've leap out of bed, switched on the light, and began the great mosquito hunt, located my prey, swatted it with a towel, then screamed with glee at the sight of its spattered body. Nowadays, I take a different and less active approach, using effective control methods coupled with natural repellents.

Control and Prevention

As with all insect pests adequate housekeeping is essential. Fit screens on all doors and windows, clean up possible breeding sites (water laying around in buckets, garden pots, etc), encourage natural predators to your garden (frogs, dragonflies, birds and fish and tadpoles in ponds), grow willow trees in wet boggy areas to drain any surplus water, and, in country areas, or anywhere that you have a watertank, fit mosquito netting to the inlet. Even permanent water, such as a creek or dam, makes an ideal place for breeding; stock it with fish to eat the mosquito larvae. You can obtain information from your local Department of Agriculture and Fisheries as to the species of freshwater fish best suited to your area. They can be purchased as fingerlings from both government and private fish hatcheries. And, of course, as the fish grow, they can be caught and eaten.

Repellents will also help to keep mosquitoes away, and are a necessity when enjoying outdoor activities.

- burning eucalyptus candles will repel them. Use them inside or at night around barbecue areas or outdoor entertainment areas. Citronella oil is also effective, and can replace eucalyptus oil in candles

- burning citronella oil in lamps outdoors at night will drive mosquitoes away
- lavender candles are also effective, and are pleasant to have burning in the home (see Candlemaking, chapter 4)
- save prunings from your herb plants, especially lavender, and throw them on your barbecue fire. Pennyroyal will also work, as will green eucalyptus leaves, the latter being used only as a last resort, since it will make a very smoky fire
- burning lavender incense inside and outdoors is another very effective repellent

Body repellents are also a must when outdoors, and the following oils, in order of their effectiveness, can be used:

lavender	*grey myrtle*
eucalyptus	*citronella*
Huon pine	*pennyroyal*
paperbark	

You can make your own fragrant oils by either enfleurage or maceration (see chapter 8) that are suitable to use as insect repellents. If you use pure essential oil, first blend 6 drops of oil with 5 ml (1 tsp) of vodka (to aid dispersal) and then dissolve the mixture in half a cup of warm water. Apply immediately.

Plants that repel
___

Castor-oil plants growing throughout the garden will help to keep it free of mosquitoes. Plant them close to outdoor entertainment areas, doors, windows, etc, but not too close, as they can grow from 2-8 metres (6'-25') tall.

Mosquito Spray
This will both kill and repel mosquitoes and breaks down quickly to leave no residual.

1 cup fresh wormwood leaves,   water and vodka sufficient
tightly packed

Place the herb in a ceramic bowl and add sufficient boiling water to just cover. Put a tea-towel over the bowl and steep overnight, strain through muslin cloth, squeezing all liquid from the herbs, and dilute with 4 parts of water to every 1 part of infusion, then add 10 ml (2 tsp) of vodka to every 500 ml (2 pt) of liquid. Store in a pump-spray bottle.

MOTHS

Clothes

It can be quite devastating at the beginning of winter when you take your blankets and woollens out of storage only to find that they have been a banquet for moth larvae. They feed on all woollen materials, including carpets and furniture fabrics.

You can use a spray to protect your clothes and other woollen items, or easily obtained camphor moth balls. However, both options are poisonous and there are safe, natural alternatives which will work just as well.

Herbs have been used for centuries as protection against clothes moths, and among the most successful are lavender, rosemary, southernwood, cotton lavender (satolina), and woodruff. Other herbs that are effective in blends are pieces of elecampane root, orris, roseroot, tansy, thyme, and spearmint. Spices can be used too — cloves, caraway seeds, nutmeg, mace, cinnamon, tonka beans and dried lemon peel. Combined and prepared as a dry pot-pourri the herbs are added to muslin or coarsely woven cotton or linen bags and placed in drawers and cupboards among woollens and other clothes. They can also be placed in the pockets of suits and jackets or hung on coathangers or wardrobe rails.

Check the scent of your moth bags every six months, and if not strong enough mix up a new batch. Use only dried herbs and ingredients.

Rosemary Moth Repellent

> *25 g (1 oz) tansy*
> *25 g (1 oz) rosemary*
> *25 g (1 oz) wormwood*

> *15 g (½ oz) freshly crushed cloves*
> *2 tbsp (2⅔ US tbsp) orrisroot powder*

Lavender Repellent

> *2 cups lavender flowers*

> *½ cup pyrethrum flowers*

1 cup rose geranium leaves | ½ cup orrisroot powder
½ cup southernwood | 2 drops oil of roses

French Moth Repellent
3 cups lavender buds | 1½ cups pyrethrum flowers
3 cups rue leaves | ½ cup orrisroot powder
1½ cups satolina (lavender cotton)

Tansy Repellent
2 cups tansy leaves | ⅓ cup ground cloves
3 cups southernwood | ½ cup orrisroot powder
2 cups thyme

You can make your own blends from any of the herbs or spices by following the pot-pourri directions (see How to Blend a Dry Pot-Pourri, chapter 4).

Pomanders

Pomanders are clove-studded oranges hung in cupboards or placed in drawers to keep them smelling fragrant and fresh. However, since moths dislike cloves they are also an effective deterrent.

To make a pomander, follow the directions (see Bedroom Herbs, Pomanders, chapter 4).

Other control methods

- wash blankets and precious woollens with the *Lavender Wool Wash and Final Rinse*
- add one or two cups of *Lavender Washing Rinse* to the final rinse cycle of your washing machine
- seal clothes or blankets in plastic bags and hang them in the sun for a day. This will kill any moths and their eggs that may have infected them
- sprinkle epsom salts through wardrobes, drawers and linen cupboards
- soak a cloth in turpentine and thoroughly rub all internal wood surfaces of drawers and wardrobes, including joints and crevices
- dried orange or lemon peel scattered in drawers and cupboards will deter moths

- vacuum woollen carpets at least weekly and shampoo on a regular basis
- sprigs of dried lavender placed at one metre (1 yd) intervals under woollen carpets will help to protect them from moth damage. When you can no longer smell the lavender it's time to replace — depending upon the amount of volatile oil contained in the herb it may be effective for as long as five years
- infected carpets can be sprayed with homemade *Pyrethrum Spray*. Test first on a small area that can't be seen to make sure that it doesn't stain. However, this type of treatment should generally not be required, as carpets are treated with an insecticide when they are made

FOOD MOTHS AND WEEVILS

There are a large number of food moths and weevils that infect our food. Most people are familiar with weevils that spoil flour or cereal-based products, and the tell-tale webbing together that betrays them.

Storing food in air-tight jars is one of the best ways of preventing infestations. However, for long term storage, or where bulk foods are opened regularly, food may still become infected. For bulk storage of grains I have found the following method effective: two-thirds fill a container that has an air-tight lid with your produce, place a candle inside and light it, replace the lid and seal around its edge with masking tape. The candle keeps burning only until all the oxygen has been used, leaving the container virtually airless. Another method I use for long-term bulk storage in extra large containers is to use dry ice — it will preserve grains, seeds, etc, indefinitely. Plastic or glass containers of 10 litre (2 gal) capacity or more, with screw-on lids are ideal. Scrub the container and lid thoroughly, and if the lid is metal give it a coat of varnish to prevent rusting.

Cover a piece of dry ice, about a quarter the size of your clenched fist, with a heavy thickness of cotton wool. This is to prevent the food to be stored from freezing. Place the wrapped ice in the bottom of the vessel, and fill to the top with your grain, etc. Lay the lid on top of the opening, but do not screw down tight. When the dry ice has melted, seal the lid tightly.

You now have the seeds in an atmosphere of inert gas in which bugs, fungi, or mould cannot survive.

Day to day storage

Use air-tight containers as already discussed in combination with the following repellents:

- try hanging small cloth sacks of black pepper in food storage cupboards
- dried bay leaves placed throughout storage jars, including flour and rice bins, is an effective moth repellent. Sassafras leaves can also be used
- put weevil repellent sachets in cupboards and amongst food. To make sachets place cloves of unpeeled garlic on a metal baking tray in a warm oven for about an hour. Combine with crushed bay leaves and add to small muslin or coarsely woven cotton bags. In large containers of food place several sachets, layering them throughout

Housekeeping

- clean up all spilled foods, especially those that are cereal or flour based. They could be potential breeding areas
- keep cupboards clean, removing all foodstuffs regularly and scrubbing thoroughly with soap and water
- dispose of all suspect foods in an outside bin. Inside bins still give eggs a chance to hatch and infect food
- wipe all internal wood surfaces of cupboards with fragrant lavender oil (see enfleurage, chapter 8). Not only will this keep your cupboards smelling fragrant, fresh and clean, it will also help to deter moths. Lavender oil will also impart natural disinfectant qualities
- for bad infestations, dispose of all infected food, empty cupboards and scrub thoroughly with soap and water, then spray all surfaces with *Feverfew Spray*. Keep well sealed for at least 3 hours, then open for another 8-10 hours. Wipe all timber surfaces with lavender oil and replace food

RODENTS — MICE AND RATS

Mice and rats are more than just a nuisance; they are pests and they spread disease.

The front line attack in controlling rodents is adequate

prevention. Install vermin wire (netting) to prevent access to under-house crawl spaces, wall cavities and roofs. Make sure that all screens fit tightly with no gaps, and that rubber strips are fitted at the base of external doors and connecting garage doors. Mice and rats can squeeze through incredibly small spaces.

Clean up areas of brush and long grass around the house and along fences and open up areas around sheds and other outbuildings. This will make rodents visible and more susceptible to attack by their natural enemies.

Rodent Control

Traps placed along runways and other haunts has always been an effective means of control. If you have rats in the house their entry is usually evident by holes chewed through walls, ceilings or doors. Finding the mousehole might not be so easy. Sprinkle baking powder in front of suspect openings at night; next morning their tracks will be seen if they have been there.

If you feel too squirmish about using traps and are not keen to use baits, the following repellents may keep rodents away.

- place fresh mint near mouseholes or a mixture of mint leaves, cayenne pepper and cotton wool soaked in peppermint oil
- mint and tansy in cupboards will repel mice
- blocks of camphor scattered around and placed in ceilings keeps mice and rats away
- rags saturated with turpentine and placed in ceilings and other rat haunts will repel them

However, once rodents have gained access to your home, no deterrent will keep them from a favourite meal. Baits provide a safe alternative to chemical poisons, and like their counterparts send the rodents away to die.

Rodent Bait

Living in a rural area, rats become a problem from time to time, especially when there is a creek and bush nearby. The following bait has worked successfully in eradicating and controlling an outbreak of rats and mice.

You will need:

½ cup cornflour   *½ cup cement*

Combine the two ingredients, mixing thoroughly, and place small amounts in margarine tubs near entrance holes or runways. For outside containers, place an ice-cream container that has had holes cut in its sides over them to provide protection from wet weather — hold in place with a large rock or half brick.

Rodents will eat the mixture, go away to find water, and the cement will react with the water, resulting in very dead pests.

If you don't wish to use cement, replace this with plaster of Paris and mix to a dough with milk.

SILVERFISH

Silverfish in the house are a nuisance and can cause irrepairable damage to books. Often the first sign of this pest is the chewed pages of a precious book. By then it is too late and the damage is done. Prevention is the best protection and should become a regular housekeeping activity.

- regular vacuuming of both books and bookshelves is essential, not forgetting books stored in boxes and cupboards
- remove books from shelves and storage at least once a year and thoroughly vacuum, including crevices and cracks in shelves, cupboards and storage boxes
- vacuum books that are to be stored away for a long time, then place in plastic bags with sprigs of lavender or cotton wool soaked in lavender oil, seal, and leave in the sun for about 6 hours
- treat baseboards, table legs and cracks in cupboards and shelves with a mixture of borax and sugar
- coarse salt sprinkled in corners and cracks is also effective. Remember, salt is hygroscopic (absorbs moisture from the atmosphere), so don't leave it too long before removing
- wipe vacuumed shelves liberally with lavender oil — this will deter silverfish for up to a year if the oil is potent
- place bunches of dried lavender in storage boxes and cupboards, or a mixture of lavender and bay leaves
- scatter bay leaves and lavender along book shelves before replacing books
- in severe cases, spray with *Pyrethrum Insecticide* (see Fleas), then adopt any of the above repellents
- and lastly, if you can bear it, do what I do and encourage

spiders, like the non-web spinning Huntsman to take up residence. Not only do they eat silverfish, but other insect pests as well

SPIDERS

Most people cringe at the thought or sight of spiders. My teenage daughter goes into a screaming fit if she sees a harmless Huntsman in her room. The reputation spiders enjoy may be unjustly deserved, since they can be a house's most valuable predator. They devour a large range of pests, including silverfish, young cockroaches, flies and mosquitoes.

People hell-bent on eradicating spiders usually justify their actions by reasoning that they are protecting their family from red-backs and funnel-webs. Without doubt, these two species are deadly, but rarely found inside the home. Funnel-webs favour moist, concealed spots, such as beneath pot-plants or rocks, while red-backs will be found mainly in rubbish heaps, brick piles, under steps, sheets of corrugated iron, window sills, old tins and drums. Storage boxes in sheds and garages are also a favourite haunt of red-backs.

Harmless spiders, however, are an asset for the control of annoying insect pests. Even the red-back becomes a tasty meal for the common black house-spider.

Funnel-webs that enter the house can be quickly despatched with a dustpan and broom or with the heel of a shoe. Red-backs can be controlled by eliminating breeding spots and regularly spraying beneath steps and window sills with homemade *Pyrethrum Insecticide*. Spray suspect funnel-web haunts as well.

If spiders inside are a definite no, brush down webs and spray with pyrethrum occasionally. Seal all cracks and crevices, install tight-fitting screens and rubber excluders on the base of external doors. Don't waste money engaging a pest control organisation; the residual sprays that they use have no continuing effect on spiders and damage the environment. Spiders need direct contact with an insecticide for it to be effective.

TERMITES

Termite damage can be disastrous and stringent control is essential, particularly in termite prone areas. Preventing them from entering the house and destroying the nest is absolutely necessary for effective control. In both circumstances professional help, prior to building a new home and as continued maintenance, is almost mandatory if you are to keep your home termite-free.

Before finalising building plans look at the options and materials so that you will not be subject to attack. Consider such things as building on concrete slabs and framing with steel and cladding with brick. Suspended floors must be well drained and ventilated, and have sufficient room to allow easy access for checking and preventive maintenance.

Preventive Measures

- fit ant caps to steps, pergolas, etc
- keep rubbish and wood piles away from the outside wall of your house
- inspect sheds, garages, etc, and around and under your house every six months
- don't store timber or timber boxes under houses where they might attract termites
- maintain a regular inspection and preventive treatment program with a reputable pest control firm

WASPS

If you have European wasps around, get rid of them quickly. You can contact your local office of the Department of Agriculture for advice, or place traps around the house and garden to eradicate them.

They are attracted by sweet things, so be careful of opened drink cans or the like in which they could be foraging. And generally avoid them, as, unlike other wasps, they are very aggressive.

However, before you decide to destroy any wasps or nests that you find, try to identify them first. Wasps, other than the European variety, usually won't bother you but will clear up an enormous number of pests, including funnel-web spiders, cabbage moth caterpillars, and pear and cherry slugs.

If you must clear away wasp nests, do it at night or during winter when they are dormant. Knock down the nest and burn the residue; if the wasps still come around, spray with homemade *Pyrethrum Spray* at night so as not to affect any friendly insects, such as bees.

EUROPEAN WASP AND FLY TRAP

(2 plastic 1.25 litre/2 pint drink bottles)

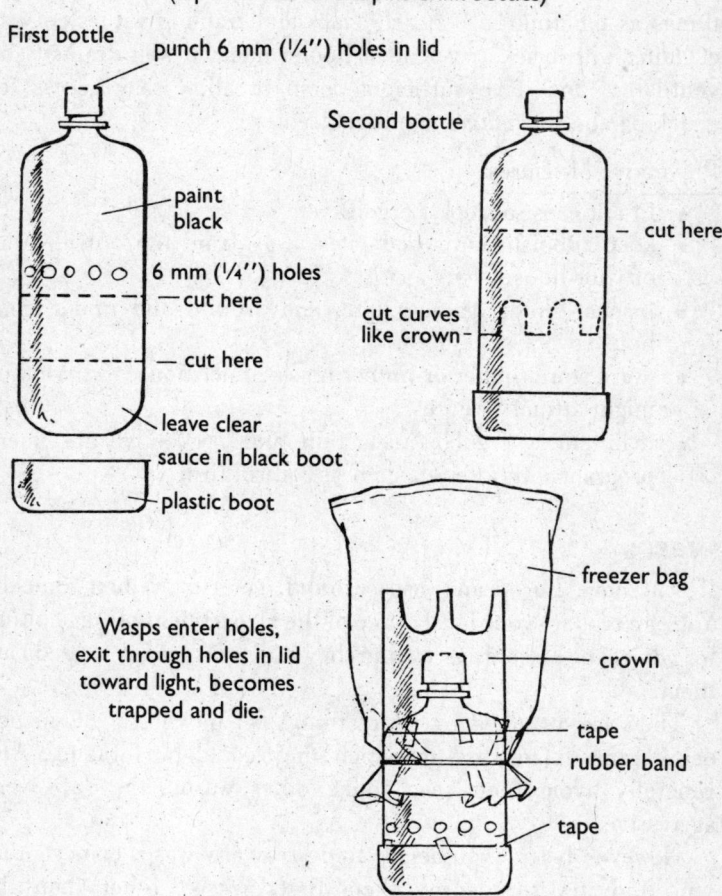

First bottle

punch 6 mm (¼″) holes in lid

Second bottle

paint black

6 mm (¼″) holes
— cut here

— cut here

cut here

cut curves like crown

— cut here

leave clear

sauce in black boot

plastic boot

Wasps enter holes, exit through holes in lid toward light, becomes trapped and die.

freezer bag

crown

tape
rubber band

tape

Put bait in saucer — honey, sugar and water. Cover saucer with net or pantyhose to prevent insects getting at bait.

7 CHEMICAL-FREE GARDENS

GOING ORGANIC

With the advent of post-modernism and the disposable-throwaway society came the onslaught of chemicals: herbicides, pesticides, insecticides and an array of fertilisers guaranteed to produce super plants. The continued, and indeed increased, production of these chemicals is without doubt a reflection of the modern attitude towards life; the demand for fast and easy results with minimum effort.

Some gardeners, persuaded by the clever marketing of the advertising agencies, reach for the latest chemical cocktail without even considering the long term consequences of constant chemical bombardment. They spray routinely as a preventive measure, zap weeds with the wave of a chemical wand, and saturate the soil with fertilisers that do nothing to improve the natural fertility and balance of the soil.

The use of these chemicals on such a large and widespread scale is not only harmful to the environment, but also to our health. Everything that is sprayed onto, or added to, the soil eventually ends up in our food chain: meat, vegetables, fruit and the water we drink. The accumulative effect of all of this is thought to be one of the main causes of severe allergies and other health problems, including cancers.

However, it is not all doom and gloom; there has now been a huge shift in attitude towards the environment, and working in harmony with it and not against it. Individuals are once again adopting the attitude of our grandparents' day: that a healthy garden demands plenty of organic matter added, and that plants should be grown in those areas best suited for their cultivation. Once again the aim is to achieve a natural balance between plants, insects and birds, so that harmony does exist.

You as an individual, if you have not already done so, can

adopt this attitude and go organic. Convert your garden from chemical dependency by composting, mulching, companion planting and practising natural pest control. It may take a while to settle your garden into a harmonious balance, but the end result is well worth waiting for.

TAKING CARE OF THE SOIL

The secret to maintaining a healthy, thriving garden is rich, viable soil. Plants will fail to thrive if soil conditions are poor, and are much more susceptible to insect attack and disease infestation.

However, before you start to garden it is important to get to know your soil by establishing its texture. This is important, since in many areas of Australia the soil is not naturally rich, and therefore may take some time to build the texture and quality needed to grow healthy plants. The three main soil categories are:

- Sandy: soil which is light and easy to cultivate. However, it has the disadvantage of losing moisture too quickly, leaving it dry and often hard-caked on the surface.
- Loam: the best type of soil to have in your garden. It is easy to cultivate, and holds moisture well. Good loam has neither too much sand nor too much clay content.
- Clay: very heavy soil and extremely hard to work. Water takes a long time to drain away.

To determine the texture of your soil, apply the following test: take half a handful of soil, lightly moisten it, form it into a sausage shape, and then gently bend it to determine its structure.

- Sandy soil will crumble and crack, and, if you try to bend it, will fall apart.
- Loam will hold together quite well and will form a shape, but will crack slightly when you try to bend it.
- Clay is smooth when moist, easy to handle and will form a shape and bend without cracking.

Once you have determined your soil type the next step is to improve its structure by the addition of plenty of organic matter, such as compost and manure on a regular basis. Mulching the surface with similar materials also improves its condition. The idea is to improve the moisture-holding capacity of the soil to more readily sustain plant life.

Clay soil requires a slightly different approach; a clay breaking substance, such as gypsum, should be added first, followed by large quantities of organic matter.

SOIL pH

The ideal soil pH level for the cultivation of the majority of plants is between 6.5 and 7.0. This can be simply tested with a strip of litmus paper, available from nurseries or hardware stores. Press the litmus paper down into slightly damp soil with your thumb. If the paper stays the same colour the soil is neutral, or 7.0 on the pH scale. Should it turn blue the soil is alkaline, and if it turns pink the soil is acid.

To correct any imbalance the following steps are necessary:

Alkaline soil: add plenty of rich organic matter, including peat and pine needles. The latter have a high acid content and as they break down help to create a balance.

Acid soil: dig in plenty of compost or well rotted animal manure, or natural lime, dolomite or wood ashes.

POOR DRAINAGE

Bad drainage can be a problem that will affect the health of some species. The simplest way to solve the problem is the inclusion of organic matter, which should improve the texture of the soil so that it will allow water to drain away more freely. Building up garden beds above ground level improves drainage, as will drainage pipes to take water away from the plant roots.

Planting deep-rooted species, such as comfrey or lupins, in badly drained spots is another alternative. When mature the plants can be chopped up with a spade and dug into the ground along with lots of well-rotted manure. In the meantime, their roots will have helped to break up the subsoil.

MINERAL DEFICIENCIES

Usually if plants are growing poorly, being attacked by insect pests, or are disease-prone, it's a fairly good indication of deficiencies. In most cases these deficiencies can be corrected by the addition of lots of organic matter in the soil, or sometimes the problem can be solved by adjusting the pH level. To identify and correct mineral deficiencies refer to the following table:

Nitrogen
slender fibrous stems, foliage and stems fade from green to yellow and the plant's growth tends to slow down.

- add plenty of well-rotted poultry manure; it has the highest nitrogen content. Other sources of nitrogen are compost, blood and bone, sewage sludge, fish scraps and cotton seed meal.

Phosphorous
growth slows down, the underside of leaves turn reddish-purple, and fruit trees may lose their fruit early or set fruit late.

- add lots of manure, compost and other organic material. Rock mineral phosphate can also be added in combination with manure.

Pottasium
plants grow poorly, leaves turn bronze and curl up, and root systems are under developed.

- make a separate compost heap that contains plenty of stable manure and green matter, plus wood ash if you can get it. Spread over the surface of the soil and keep well mulched with organic matter.

Boron
slow growth, plants become bushy and terminal buds die. Later lateral buds die, roots or tubers crack and leaves thicken.

- mulch well with well-rotted manure or compost.

Calcium
thick woody stems and retarded growth, and blackening and dying of roots. In some instances terminal buds and young branches may deform.

- treat with ground natural limestone.

Copper
slow growth; shoots and tips die back — this is especially evident in fruit trees.

- apply plenty of compost or well-rotted manure; adding rock minerals is helpful and can be included in the compost heap.

Iron
leaves on top of the plant turn yellow, and spotted, coloured areas are evident on young foliage.

- avoid over-liming; add plenty of well-rotted manure, crop residues, and blood and bone.

Magnesium
growth is very slow and plants are late to mature. Entire leaf can become mottled with dead areas, and the spaces between veins on older leaves may turn yellow. On fruit trees watch for patches of dead tissue on older leaves, or leaves dropping — first on old branches, then on twigs of current season.

- add seaweed to your compost heap, or a litre (2 pt) of sea water to every 50 kilos (110 lb) of compost. Incorporate dolomite directly into the soil or include in your compost pile.

Zinc
leaves may become mottled and turn yellow, developing an abnormally long and narrow shape.

- can be a common problem in peat soils, and requires the incorporation of plenty of well-rotted manure and raw phosphate rock minerals, if you can get it.

NURTURING THE SOIL
Good soil is the result of continual nurturing and building up with organic matter. Regardless of whether your soil is sandy or clay, or deficient in minerals, the addition of compost and manure will result in a viable soil and make gardening easier and the results rewarding.

Composting is usually the best starting point for organic gardening, as it is an environmentally sound means of recycling garden and food wastes. It also provides one of the most basic means of conservation, assuring that future generations will have the same benefit of the earth as we do now.

Composting
The principle of composting is simple, and each individual should see it as an essential part of a responsible and efficient home management system: if it once lived, when you're finished with it, return it to the earth where it belongs.

Composting is, in broadest terms, the biological reduction of organic wastes to humus: a rich humus with a slightly sweet, earthy aroma.

A healthy heap needs both water and air to generate the heat required to hasten the breaking-down process. If allowed to dry out the process will slow, likewise, if air cannot circulate the entire process will take much longer and may even stop.

Effective compost should be built layer upon layer, alternating food wastes with an activator such as animal manure, blood and bone, or liquid seaweed. The nitrogen and protein content of the activator accelerates the breakdown of the organic matter and encourages the bacteria which heat up the heap.

Useful composting ingredients include:

- leaves — layer them in the compost or make leaf mulch
- grass clippings — ideal for covering kitchen scraps. They are fast working and create a lot of heat that helps to hasten the decomposition of other ingredients
- seaweed — rinse salt off and chop before adding to the heap. It has the advantage of rotting quickly and containing many useful minerals
- rotted sawdust — speeds up decomposition
- wood ash — rich in potassium
- shredded paper — to 10 per cent of total
- weeds — providing your heap is hot and quick to decompose, weed seeds will be destroyed
- straw — provides bulk; is best added as stable sweepings as a mixture of straw and manure. Cover with grass clippings
- kitchen wastes — vegetable scraps and peels, fruit skins, melon rinds, apple cores, nuts, shells, tea leaves, coffee grounds and crushed egg shells (these will add calcium)
- vacuum cleaner bag content and floor sweepings

To make your compost heap

- clear a patch of ground, removing grass and levelling if necessary. Compost should always be built on soil and never on concrete
- place a few bricks down on edge to allow air to circulate into the heap
- put down your first layer of material — grass clippings,

garden wastes, kitchen scraps, etc, in the middle
- next dust over a layer of fowl or cow manure, dolomite or blood and bone to a depth of about 1 cm (⅜")
- sprinkle with water. Repeat the procedure until your pile is built
- after a week, turn the pile over with a fork to speed decomposition

A compost pile about 1 metre (3') in height should be broken down into humus after two months in summer, but longer in winter. Add to garden soil in spring and autumn at the rate of 1 kg (2 lb) per square metre (yard), or a 5 cm (2") covering over the garden bed. It can also be used for container plants, raising seedlings and mulching around growing plants.

Compost Bins

Compost can easily be made in a heap on the ground, building it up from the sides and inwards, or contained within some type of formal structure. It can be made from timber slats, bush poles, concrete blocks, bricks or corrugated iron. The bin should consist of an earth floor, be open on one side, and have holes for air circulation in the other three sides.

To keep the contents from spilling out, the open side can be fitted with a removable piece of welded-mesh wire.

There are a number of different commercially manufactured bins and tumblers available. However, the compost they make is no better than a homemade structure and it involves an unnecessary expenditure.

Underground Composting

If you don't have a compost bin or pile, simply dig a hole in the garden, fill it with kitchen scraps, sprinkle a cup of dolomite over it, and cover again with soil. After about a week earthworms will have the soil workable, giving you a high-quality humus.

Herbal Compost Activator
A bio-dynamic activator that will give you a quick return compost.

1 part stinging nettle	1 part chamomile
1 part yarrow	1 part valerian
1 part dandelion	1 part oak bark

Mix dried and crushed herbs together thoroughly, and then add 1 part of the mixture to a lidded bucket containing 20 parts of rain water. Shake well and leave to stand for 24 hours.

With a large stick (about 50 mm/2″ diameter) poke holes about every 30 cm (1′) in your compost heap. Pour 75 ml (2½ fl oz) of activator in each hole, cover with soil and leave for one month, after which time the compost should be rich, black and crumbly. Ideal for heaps of tough fibrous material, soft weeds, grass and manure.

When starting your heap, let it settle for 2-3 days before adding the activator.

Natural Fertilisers

Compost Tea

Since plants drink their food rather than eat it, the use of compost tea makes sense, particularly during dry weather when plants are starved both for food and water. Some of the nutrients in compost dissolve quickly and readily in water, and in solution these nutrients can be quickly distributed to needy plant roots.

To make the tea, add about 10 cm (4″) of mature compost to a bucket of water. Allow to stand for 48 hours, stirring occasionally, then strain through coarsely woven cloth.

Apply undiluted to plant roots or use as a foliar feeder by spraying or sprinkling over the leaves of plants. It is especially good for leafy vegetables, such as silver beet, spinach, cabbage and lettuce.

Mix a handful of woodash to the solution before using and it can double as a natural insecticide.

Manures

The reason for manure being such a good, balanced fertiliser is that up to 80 per cent of all nutrients gathered up by farm animals in foraging is expelled in their dung. However, the amount of nutrient in these manures, even when dry, is small compared to chemical fertilisers. Therefore it may take two tonnes of cow manure to do the same job as 50 kilograms (110 lb) of ammonium sulphate. But, most importantly, chemicals do not supply humus, or a well balanced food supply for plants or earthworms, which the animal manure does.

Cattle manure is a compact type of cake that breaks down slowly. This makes it long lasting, a desirable characteristic for feeding the soil as a mulch. By comparison, the loose textured nature of horse manure is unreliable as a source of nutrient when left in the open.

Poultry manure is higher in potassium than any of the other manures. However, its humus value is poor unless mixed with a deep litter of sawdust, straw, or other fibrous material. Yet its popularity lies in the fact that, unlike cow or horse manure, it's not likely to introduce weed seeds into the garden or compost.

There is one drawback to poultry manure: unless it is mixed with some form of fibrous material its nitrogen content, in the form of ammonium carbonate, is fierce and can burn crops harshly.

Sheep manure is another valuable and rich source of nitrogen, especially if it can be obtained fresh in an unleached condition. If you can locate a readily available source it makes excellent manure tea, or can be used under a mulch of grass clippings around trees and plants.

Horse manure produces a great deal of heat during its rapid decomposition, and is therefore an ideal activator in the compost heap. It can also be dug successfully into heavy soils to lighten them.

By far the richest manure of all is pigeon droppings, with four times more potash and nitrogen than poultry manure. The phosphorus content is also double, making it an ideal choice as a compost activator, because it is not safe to dig directly into a highly productive garden.

Green Manuring

Deficient soils can be improved by planting green manure crops, such as lupins and cowpeas. They benefit the soil by:

- improving the physical condition or crumb structure of the soil
- conserving mineral matter and increasing nitrogen content
- improving drainage
- temporarily increasing the quantity of organic matter

A green manure crop is a means of providing a source of readily available plant food, which benefits succeeding crops by the fertilising ingredients contained in its decaying mass of vegetation.

The process simply involves growing the appropriate crop, allowing top growth to develop, slashing it down and then digging it into the soil to rot. However, its success depends upon conditions favourable to the decomposition of the buried material: sufficient warmth and moisture. The more immature the crop, and the lighter textured and aerated the soil, the more rapidly the nutrients are released.

Green manure crops can be sown at any time provided there is ample water. Though it is best to grow crops suited to their season, New Zealand blue lupins, field peas or vetches can be sown in autumn, and in spring and early summer cowpeas and Japanese millet, alone or in combination, are good green manure crops.

Broadcast the seed at the following rate per 8 square metres (9½ sq yd):

- lupins and field peas: 200 g (7 oz)
- vetches, cowpeas and Japanese millet: 125 g (4½ oz)

Sow the seed on prepared moist soil and cover to a depth of 2-5 cm (1"-2").

After digging in, the next crop may be commenced immediately, taking care not to place seeds or seedlings in a pocket of chopped up material.

Wood Ash

As a natural fertiliser, wood ash varies considerably in composition. Generally, it does not contain more than 2-3 per cent of potassium, but does include appreciable amounts of lime and magnesia and some phosphate.

It does have some fertiliser value, but should be added only to the compost pile and not applied directly to the soil.

Leaf Mould

Bush leaf mould is a very poor source of organic matter if dug directly into the soil. It seriously depletes the amount of available nitrogen, resulting in a marked reduction of plant vigour. However, it can be used after having rotted in the compost heap.

Seaweed

Seaweed should only be used where it can be gathered locally, as it is not worth paying cartage. It contains a reasonable amount of potash, and also small quantities of iodine and boron. If chopped up finely it can be used in the compost heap or as a mulch on top of the soil.

Steeped in water alone, or in combination with manure, it can be applied at soil level or sprayed onto the foliage of plants. Liquid seaweed is also a useful fungicide for vegetables, ornamentals and trees.

Liquid Fertilisers

Liquid fertilisers can be simply made using readily available natural materials.

Liquid Manure

This is best made in a clean 200 litre (40 gallon) drum (with one end removed) or a large galvanised garbage bin.

To make good liquid manure, place animal manure of one type, and equal to one-fifth of the capacity of your container, in a hessian bag and suspend it in a container of water. Cover the container to keep flies away and leave the manure to soak for 2-3 weeks, stirring occasionally. This will give you a potent liquid fertiliser concentrate that will give your plants that extra boost, and is a particularly excellent stimulant for green vegetables such as silver beet, lettuce, cabbage and cauliflowers.

Don't apply the fertiliser in concentrate form, because too much of a good thing is just as bad as too little. Instead dilute half a litre/1 pint (no more) to a bucket of water and apply to the garden. (It should have the colour of strong tea.) Repeat this operation every two weeks for flowers and vegetables and monthly for shrubs.

It is best to apply after rain or watering the garden and never during a drought without first giving your plants a good watering. Liquid manure can also be added to the compost as an activator or used to prepare dug-over ground for next season's vegetables.

When your concentrate has been exhausted, top up your container with water and again steep for 2-3 weeks. This time add one litre (2 pt) of fertiliser to a bucket of water. You can steep the manure a third time, using one-and-a-half litres (3 pt) of liquid to a bucket of water. Once the last batch has been exhausted, add the manure to your compost and start afresh.

Liquid Comfrey

Comfrey is a green leafy herb available from most nurseries and easily grown in the backyard garden. Its leaves make a thick green, pungent liquid manure that can be used for all garden plants, and is especially good for promoting rapid growth in vegetables.

To make the fertiliser, pick fresh comfrey leaves and half-fill a large bucket, preferably one with a lid, or a large plastic garbage bin, then fill with water. Leave to steep for 3 months and then dilute one part comfrey to two parts water before sprinkling around plants.

Couch Grass

Couch grass makes a good liquid fertiliser and is made and used the same as liquid comfrey.

Nettle Fertiliser

This is a nitrogen-rich liquid fertiliser that can be used for all plants, and is especially good for green leaf vegetables. However, only apply to beans when they first emerge and when flowers are set. Otherwise the plant will produce more leaf in favour of pods.

Rot a large quantity of stinging nettle leaves in a bucket or drum filled with water. Leave to steep for two weeks and dilute one part nettle liquid to ten parts water.

Use as a liquid plant food every fortnight during the growing season.

Liquid Seaweed

Gather sufficient seaweed to one-fifth fill a drum or large garbage bin, rinsing away thoroughly all traces of salt before using. Cover with water and allow to steep for three weeks, then dilute with two parts of fresh water. Spray onto the foliage of plants or apply at soil level.

When the concentrate is exhausted the seaweed can be used as a mulch or added to the compost.

Diluted Urine

If it doesn't worry you to collect, human urine is a good source of nitrogen and your plants will love it, provided it is diluted one part urine to five parts water. It is also essential that everyone in the family is in good health and not suffering from any major medical problem, such as hepatitis.

Apply to soil around plants once or twice a year using a watering can, or add to the compost as an activator.

Bio-dynamic Fertiliser

An excellent liquid fertiliser that guarantees healthy vegetable crops. Spray the following solution onto garden soil every two weeks during autumn.

1 tbsp (1¹/₃ US tbsp) dried powdered cow manure

2 tbsp dried powdered seaweed

1 cup dried Fat Hen leaves (Chenopodium album — a wild herb that grows prolifically throughout Australia, with the exception of the colder regions)

1 cup dried dandelion leaves

1 cup dried stinging nettle leaves

1 cup purslane leaves

1 cup dried chamomile flowers

1 cup dried sage leaves

Add the powdered manure and seaweed to 4.5 litres (1 gal) of rain water and stir until dissolved. Cover the container with clear plastic and leave to soak in the sun for three weeks.

Reduce all the dried herbs to a powder by rubbing through a fine wire sieve.

Strain 1 cup of the manure/seaweed liquid and add to a drum containing 25 litres (6½ gal) of rain water. Add powdered herbs, stirring well, cover with clear plastic and leave in the sun for two days before using.

Mulching

Mulching your garden with organic matter is one of the best and most efficient ways of keeping soil and plants healthy and happy. They generally improve soil texture as well as providing a number of other benefits.

- it prevents weeds from emerging in the soil between rows and around the base of plants
- it reduces the need for excessive watering by keeping the soil moist, and prevents loss of moisture through evaporation

- a steady stream of nutrients is supplied to the plant roots
- soil texture is improved because it's kept lightly moist
- decaying mulch builds the soil into rich, friable humus

Apply the mulch as it breaks down, maintaining a good deep layer. However, be sure that the mulch layer is not placed too close to the base of newly planted seedlings.

You can also layer your mulch with different materials to cater for specific needs: a layer of blood and bone at ground level to provide immediate nutrients; then well-rotted and chopped manure to improve soil texture, followed by a thick layer of grass clippings.

Materials suitable for mulching are:

- mushroom compost, if a source is handy
- homemade compost, suitable for all soil types and conditions
- manure — cow or horse are the best choice and should be pulverised before applying to the soil. This can be done by running the lawnmower over it a few times until sufficiently broken down. Well rotted poultry manure can also be used and should be mixed with some type of fibrous litter, and its high nitrogen content is excellent for promoting leaf growth in vegetables such as silver beet, lettuce, etc
- grass clippings — use as a top layer mulch only unless combined with manure or compost
- pine bark chips — excellent top layer mulch or to use in combination with manure or compost
- leaf mould or mulch — can only be used after is has become well rotted. Either add in layers to the compost or rake up leaves into a pile in autumn, water well and allow to break down friable mulch.

Watering

Plants need moist soil so that they can take up the various nutrients essential for their growth. However, water is a precious resource that should never be wasted or taken for granted. Saving it and reducing its use can be achieved effectively by mulching the soil and adopting appropriate watering methods — these are discussed in chapter 1.

Other points to consider are:

- begin a deep watering regime: water roots and not leaves to encourage roots to travel downwards in search of moisture, as this helps to establish plant self-sufficiency so that gardens require less watering during dry periods
- avoid establishing beds at the base of trees where there will be competition for water
- incorporate, if possible, hedges or windbreaks to reduce the effect of prevailing winds. You will create a micro-climate that will require less watering
- watch for heat stress in summer and water those species accordingly
- in summer, water early morning, and in winter, or cold climates, during mid-morning
- consider installing a rainwater tank (see Conservation, Water, chapter 1).

Earthworms

Garden soil that is alive with earthworms is rich and productive, and each time I look at my own patch the healthy, strong plants and vegetables bear testimony to this. Each time I break open the soil to set out a young seedling it's a mass of waving fingers, some of them even as thick as 5 mm (½″).

Soil depleted by overuse, excessive chemical abuse, or lacking in organic material will contain very few if any earthworms. Their addition to the soil will not instantly give you a rich, friable humus, but they will certainly assist in restoring a balance. However, combined with added compost, manure and organic fertilisers the earthworms will rapidly multiply, gradually improving your soil.

Companion Planting

Certain plants grow harmoniously side by side, because different plants give off substances through their leaves and roots which compliment the needs of their neighbour. Some of these compatibilities have been proven to be physical or chemical reactions — tall growing plants like corn will protect shade-loving plants such as watermelon and cucumbers, while marigolds give off a substance through their roots which repels nematodes (eel worms). Other plants, for instance garlic, will repel flying insects and all plants growing near thyme will be invigorated by it. Nasturtiums wandering among vegetables will drive away aphids and keep the cucumber beetle at bay.

In nature plant species do not grow close together in rows but happen spontaneously. This companion growing, explained in scientific terms, is called 'natural selection', and can be applied to the home garden. Ail that it requires is a little thought, and the results will be basically the same as Mother Nature intended. Quite simply it works like this:

- shallow rooted plants thrive when growing near deep rooted species
- leafy crops are best planted with root crops — this way they are not competing for the same ground space or specific nutrients
- tall species will provide shade and shelter for smaller shade-loving plants
- species with different water requirements do well together as they are not competing with each other. This equally applies with plants of different nutrient requirements
- some species, such as peas, have roots which give off nitrogen, which will benefit those plants that need plenty of nitrogen
- other species, like those already mentioned, exude aromas that discourage insects

Companion planting helps to create a healthy garden, which in turn is less vulnerable to disease and insect attack. It allows you to work in harmony with nature and not against it.

Species that grow well together or repel insects:

Basil	repels flies and mosquitoes, and white fly from tomato
Beans	will thrive when planted beside carrots, cauliflowers, cucumbers, corn and radishes. They dislike onions, beetroot, garlic, gladioli and sunflowers. Summer savory is a friend that repels bean beetles
Beetroot	will do well with any plants as companions except pole (climbing) beans,

which will stunt their growth. They do exceptionally well when planted with or near lettuce, cabbage, onions and bush beans

Cabbage

strawberries, pole beans, or tomatoes close by will discourage cabbage growth. Rosemary, mint, thyme and other aromatic herbs repel the cabbage worm and adult butterfly. Sage gives off camphor, which also repels the cabbage moth

Carrots

dislike only dill, which is a relative. Carrots grow well with Brussels sprouts, peas, cabbage, lettuce, radishes and chives. Leeks, sage and rosemary repel the carrot fly. Plant carrots and leeks in alternate rows, as the carrots will repel the onion fly

Catnip

flea beetles are repelled, while cats are attracted by it

Cauliflower

aromatic herbs benefit cauliflower

Chives

will enhance the growth of carrots and tomatoes while repelling a whole host of insects

Corn

the addition of nitrogen to the soil by peas and beans enhances the growth of corn. Corn itself stimulates cucumbers, melons and squash, and grown near tomatoes will lure away the pest heliothis

Cucumbers

one of the few vegetables that dislike aromatic herbs. Likewise, they do not do well with potatoes, but instead prefer corn, cabbage, bush beans, lettuce or radishes. Radishes keep the cucumber beetle away from all cucumber relatives

Garlic

its smell is a repellent to Japanese beetles, aphids, mosquitoes, caterpillars and sucking bugs. Its aromatic oil will also control some blight diseases

Lemon balm

attracts bees to the garden and can be used as a border edging in combination with marigolds

Lettuce

is aided by root crops, especially beetroot and carrots. Onions and cabbage are helped along by lettuce growing with them

Marigolds

the foliage is reputed to repel bean beetles, tomato fruit worms, flea beetles and white fly. The roots exude a substance which will kill nematodes in the soil — however, to be effective in the soil, it must be grown in the same spot for a few years

Marjoram

this herb is credited with stimulating almost everything in the garden

Melons

corn and melons are good companions

Mint

repels the cabbage butterfly and ants

Nasturtiums

will repel aphids, cucumber beetles and give radishes a great hot flavour

Onions

oil from onions inhibits the growth of beans and peas. Onions like most other plants, especially beetroot, cabbage, lettuce and summer savory. When planted with carrots will repel the carrot fly, and likewise the onion fly will be inhibited by the carrots

Oregano

similar to marjoram, oregano stimulates plant growth in general

Peas

dislike onions, garlic and potatoes. The nitrogen that peas provide the soil benefits most other plants. Carrots give off a root exudate that benefits peas. They also grow better around cucumbers, corn and radishes

Peppers

onions, carrots, tomatoes and eggplant are good companions

Radishes

these plants appear to have no enemies and favour the presence of carrots, lettuce, peas, cucumbers and climbing peas. Plant nasturtiums close by for a better radish flavour

Rosemary

will repel cabbage moth, bean beetles and carrot fly

Rue

makes a useful edging hedge that insects won't go near, and snails give it a wide berth. Keep away from sage and basil; they don't mix

Sage

the camphor given off by this plant repels cabbage butterfly, bean beetle and carrot fly

Squash

grow well with corn

Stinging Nettle

grown near tomatoes will protect them from mould

Summer Savory

planted between rows of beans will inhibit the bean beetle — onions will appreciate it, too

Thyme

repels cabbage worm, and all plants growing near it are invigorated

Tomatoes

grows well near onions, asparagus, parsley, celery, basil, carrots and chives. Never grow near potatoes, dill or cabbages

Crop Rotation

Rotating crops every season has been practised by organic gardeners for centuries. It guarantees the best results from the vegetable garden, ensuring that there is not a build-up of soil-borne diseases that attack a particular crop and that the different needs of crops are catered for: nutrient deficiencies are prevented, as different plant groups take different nutrients from the soil.

To practise rotation, divide your vegetable garden into thirds and plant out as follows:

	PLOT A	PLOT B	PLOT C
1st Year	roots	legumes and other crops	brassicas
2nd Year	brassicas	roots	legumes and other crops
3rd Year	legumes and other crops	brassicas	roots

Root crops include:	beetroot, carrot, chicory, parsnip, potato, silver beet, swede and turnips
Legumes and other crops include:	beans, peas, celery, leeks, the various onions, sweetcorn, squash, cucumbers, melons, pumpkins, zucchini, peppers, tomatoes, and eggplant
Brassicas include:	broccoli, Brussels sprouts, cabbage, cauliflower, kale and kohl rabi

Salad crops not mentioned in these groups, including radishes, can be grown as catch crops (between the rows of longer-maturing vegetables) or as a succession crop: between the harvesting and sowing of a major crop. Salad vegetables can also be allocated their own plot close to the kitchen.

The perennials, such as rhubarb and asparagus, should be allocated a permanent space as they will crop in the same site for a number of years.

If your vegetable garden space is large enough you can adopt the following rotation method, which is the system I now use. As previously, divide your plot into three and plant out as follows:

	PLOT A	PLOT B	PLOT C
1st Year	all vegetables	green manure crop	left fallow — organic matter added

	PLOT A	PLOT B	PLOT C
2nd Year	left fallow — organic matter added	all vegetables	green manure crop
3rd Year	green manure crop	left fallow — organic matter added	all vegetables

The vegetable plot is planted out each year according to the companion planting guide in this chapter. This system allows the fixing of nitrogen in the soil and helps to prevent the build-up of soil-borne disease by leaving one section fallow each year.

Other factors to consider before you start:

- pH requirements vary from plant to plant. Avoid positioning acid-loving plants in ground that has been limed
- follow a heavy-feeding crop with one that is far less demanding
- leafy crops need lots of nitrogen, while root crops feed heavily on potash. Rotation between the two will achieve a balance

A No-Dig Garden

This type of organic garden is ideally suited for areas with poor or rocky ground where digging and cultivating would be difficult. A sloping patch of ground on my own property, which was clay with little topsoil, was terraced and then improved by this method — it now produces bumper vegetable crops.

The principle is simple: different layers of organic matter are built up on top of each other from ground level. These layers are never disturbed by any form of cultivation, but simply added to as they decompose and mulch down. To start your no-dig garden add a layer of sawdust at ground level, followed by a layer of seaweed, folded newspapers, and then alternating layers of straw (if available), grass clippings, manure and compost. A thick layer of kitchen scraps can also be added on top of the newspapers, if desired.

Seeds or seedlings are planted in the top layers of organic material, and then watered and fed as usual, with more organic material being added each season as a mulch. By continually adding layers of organic matter, good drainage is achieved and plants will thrive in a healthy environment.

Controlling Pests and Diseases

The most logical approach is prevention: a healthy, well-grown garden will be less vulnerable to pests and diseases. This can be achieved by adhering to the following simple guidelines:

- species selection — grow only those plants suited to the climate in which you live
- good soil — maintain pH level, add plenty of organic matter and ensure adequate drainage, all of which have already been discussed in this chapter
- regular maintenance — always mulch (manure will help provide a steady stream of nutrients), water well, especially during summer, feed as required, and keep the garden free of weeds
- plant placement — grow species according to their needs; don't plant shade loving varieties in full sun and vice-versa
- overcrowding — plants too close together may grow weak and subsequently be prone to insect and disease attack
- hygiene — keep garden free of weeds, remove fallen fruit from the base of fruit trees, and remove any diseased plants
- crop rotation — practise rotation of vegetables as already outlined
- companion planting — follow the guide in this chapter
- biological control — organic methods will maintain a healthy garden environment, encouraging birds and friendly insects and other creatures, such as skink lizards, into your garden — they will devour many of the harmful pests

Natural Insect Control

- each morning (if possible) inspect your plants and carefully pick off small insects like aphids and caterpillars by hand
- badly attacked plants should be completely removed and in most cases burnt. The latter helps to prevent the problem spreading
- try sponging with warm soapy water to remove aphids and other leaf-sucking insects, or apply a natural organic spray.

Repeat the treatment every few days until the problem disappears

- birds can be discouraged from destroying a fruit crop by hanging aluminium reflectors amongst branches or covering the tree with a net
- fruit fly can be controlled by the use of homemade traps hung in fruit trees or near other plants usually affected by them. To make a trap see Household Pests, Wasps, chapter 6, or pierce the centre third of a 1.25 litre (2 pt) plastic drink bottle with 5 mm (¼″) holes, add a mixture of ammonia, water and sugar to a depth of 2 cm (¾″), and replace the lid. This will attract and kill them
- snails and slugs can be discouraged by using a natural bait or by placing barriers in their path (see Insect Barriers, this chapter)
- the foliage of brassicas (broccoli, Brussels sprouts, cabbage, cauliflower, kale and kohl rabi) can be sprayed with salty water to prevent cabbage white butterfly from laying eggs
- make your own organic insect sprays and fungicides from herbs and other plants

Fungicides

Chamomile Tea Spray

Use to help destroy damping off fungus and powdery and downy mildew.

| 1 cup dried chamomile flowers | 3 cups boiling water |

Add the herb to a ceramic bowl, pour in boiling water, cover and steep overnight. Strain through muslin cloth, squeezing all liquid from the herbs, and spray onto affected plants.

Chive Spray

This is an effective fungicide to combat mildews that attack members of the squash family, including zucchini, and can also be used against apple scab.

| 1 cup freshly chopped chives | 3 cups boiling water |

Prepare as for chamomile tea spray, infusing for 15 minutes. Use undiluted, spraying onto foliage.

Seaweed Spray

A useful fungicide to help control mildew, brown rot, curly leaf and other fungi on vegetables, ornamentals and trees. Prepare as for liquid seaweed (see Natural Fertilisers, this chapter), and spray on affected plants as required.

Elder Spray

A strong spray for mildews, black spot and a variety of fungi, and is also poisonous. Store out of the reach of children in a well sealed bottle in a dark, dry cupboard for up to three months. Clearly mark the bottle 'POISONOUS'.

> 500 g (17¾ oz) elder leaves   water sufficient to make
> 1 litre (2 pt)

Place the elder leaves in an enamel saucepan, add 1 litre (2 pt) of water, cover with a lid and boil for one hour. Top up if necessary. Strain through muslin cloth when cool, squeezing all liquid from the herbs, add water to make up 1 litre (2 pt), bottle and label.

Use as required, spraying onto affected plants.

Horseradish Spray

A general purpose spray for any fungal problems.

> 1 cup tightly packed, freshly   750 ml (25 fl oz) boiling
> chopped horseradish leaves   water

Prepare as for chamomile spray, and apply undiluted to affected plants.

Mildew Dusting Powder

An effective powder to use in the control of powdery mildew.

> mustard seeds, sufficient

Using a pestle and mortar, grind the mustard seeds to a fine powder.

Store in a plastic bottle with holes punched in the lid. Dust on affected plants as required until problem is under control.

Bordeaux Paste

This is an organic substance traditionally used for treating tree wounds and collar rot.

> 45 g (1½ oz) copper sulphate   90 g (3 oz) brickie's lime

1½ l (3 pt) water

1 tbsp (1⅓ US tbsp)
powdered skim milk

dissolved in 1½ l (3 pt) of
water

Dissolve the copper sulphate and skim milk in water and then thoroughly mix with lime/water solution. Use as required.

Insecticides

Nasturtium Spray
Use this to keep woolly aphids away from fruit trees.

2 cups tightly packed
nasturtium leaves

1½ l (3 pt) water

In an enamel saucepan, bring the herbs and water to the boil and then simmer for 15 minutes. Steep until cool, then strain through muslin cloth, squeezing all liquid from the herbs. Spray as needed.

Nasturtiums grown at the base of fruit trees will also help to keep these pests away.

Garlic Insecticide
Garlic is a general insecticide spray for use with most insects, including aphids and caterpillars.

1 cup chopped garlic cloves

40 ml (1⅓ fl oz) mineral oil

½ l (1 pt) water

soapy water sufficient

In an air-tight jar soak the garlic in the mineral oil for two days. Add ½ litre (1 pt) of water, mix well, strain and stir in ½ litre (1 pt) of soapy water.

To make the soapy water, dissolve 10 g (⅓ oz) soft soap (available from chemists) in ½ litre (1 pt) of hot water. Allow to cool before mixing with garlic solution. Soft soap is based on potassium carbonate instead of caustic soda.

Before use, dilute one part garlic concentrate with 49 parts water and spray as required. It will cover an approximate area of 25 square metres (30 sq yd) if the total amount of concentrate is diluted.

Garlic-Pepper Spray
This is stronger than the garlic spray and helps to combat most insects, scale and hard-surfaced pests such as shield bugs.

Prepare as you would the garlic spray, including four chopped hot chillies steeped with the garlic in the mineral oil.

Quassia Spray
Effective against aphids and small caterpillars.

> 60 g (2 oz) quassia chips   2½ l (5 pt) water
> (available from the chemist)

Combine the quassia chips and water in an enamel pan and boil for two hours. Strain and dilute 4:1 with water before using.

Rhubarb Spray
Spray on fruit trees to discourage aphids. The leaves of the plant are poisonous, so mark the bottle clearly and keep out of the reach of children. However, it does break down quickly and is harmless to bees.

> 1½ rhubarb leaves   3 l (6 pt) water

Boil the leaves in water for half an hour, strain, and add 28 g (1 oz) soft soap. Dilute 1 part solution to 1 part water before using.

Turnip Spray
This insecticide will kill red spider mites and pea aphids, and is harmless to humans or animals.

> 1 kg (2 lb) chopped turnip   water sufficient
> root

Put the roots in an enamel saucepan, add sufficient water to cover plus 1 cm (⅜") gently boil for 30 minutes. Remove from heat, keep covered, steep overnight, and strain. Dilute with an equal quantity of water before use.

Soap and Washing Soda
Use to control leaf sucking insects such as aphids and scale. Store in a tightly sealed bottle until needed.

> 100 g (3½ oz) soft soap   water sufficient
> 250 g (9 oz) washing soda

Mix the washing soda with a small quantity of very hot water, stir in the soft soap until dissolved, and dilute with 10 litres of cold water.

Chilli Dusting Powder

This can be used to dust cabbages, cauliflowers, etc, and tomato plants to kill caterpillars.

Wearing rubber gloves, grind up sufficient hot chillies in a pestle and mortar until reduced to a powder. Store in a suitable container with holes in its lid. Use as required.

Be careful not to touch your face, mouth or eyes when handling the chillies — it burns and irritates for quite a long time.

Pepper Spray

Good for most garden pests and can be sprayed on or poured over plants. Equal quantities of:

fresh spearmint leaves	*horseradish roots and leaves*
green onion tops	*hot chilli peppers*

Combine all the ingredients with a little water and run through a blender. Add this to 4 litres (1 gal) of soapy water, and dilute before use by adding 125 ml (4 fl oz) of the mixture to 1 litre (2 pt) of plain water.

Insect Barriers

- to prevent the larvae of codlin moth crawling up the trunk of trees and attacking fruit:
  - apply a band of grease around the tree trunk: mix 4 parts powdered resin, 2 parts turps, 2 parts linseed oil and ½ part honey. Combine all the ingredients in an old pot, bring to the boil and simmer for 15 minutes. Apply to trees while still warm — any left over grease can be stored for up to 12 months. Simply re-heat and re-use. I have found this to be an extremely effective method in controlling this pest in my own fruit trees.
- snails, slugs and leaf eating insects can be discouraged in the following ways:
  - barriers such as sawdust, crushed egg shells, sand or wood ashes will all deter snails and slugs, as they dislike crossing the coarse textures
  - small circular rounds of mosquito netting placed over young seedlings will keep snails from them. Remove once plants are grown and established

— mix equal parts of lime, wood ash and bran together and sprinkle around the edges of garden beds. This is an excellent snail barrier

— salt sprinkled around young plants and the garden will make short work of snails and slugs. Salt can also be sprinkled on any of the other barriers

— a slug and snail trap can be made by sinking small dishes of stale beer and sweetened water in the garden at ground level

— those plants susceptible to attack by white cabbage moth can be protected with mosquito netting. Make small semi-circular frames from no. 8 fencing wire and attach mosquito netting — place this over young seedlings until they are large enough to survive an attack. Once plants are fully established the white moth caterpillars will only eat the outer leaves

CHICKENS IN THE BACKYARD

They're not everyone's idea of bliss, and in some areas the local council won't permit them. However, if you have a large enough yard, the council says yes, and neighbours won't mind, consider running 3 or 4 chooks.

The most efficient chicken run for the suburbs is the moveable ark or A-frame, with one end covered for perches and laying boxes and the rest covered with wire-netting. My father used this system for many years to weed and fertilise his vegetable garden. The run has no base and is lifted and moved around as desired.

Laying boxes are built against the covered end wall with outside hatches for access. Once the chickens have handed over their daily bounty you can let them out to scratch around for a while - they love to look for worms in the compost, and will be turning it over at the same time. However, make sure they don't have access to young seedlings or even the vegetable patch — this could end up in disaster.

Free range chickens will only need organic wheat and a small amount of mash each day, and plenty of fresh water and shade in summer. Keep the cage clean and grow a few pots of southernwood next to it and you won't be bothered with lice and other pests.

Buy only healthy stock from a reputable dealer, and only chickens that haven't been debeaked. Pullets are usually sold when 3 months old at point of lay. Check your telephone book for a free-range chicken farm; the birds will be healthier than the mass-produced variety. Leghorns always seem to be the most popular choice and are good layers, but you may prefer the coloured varieties such as Rhode Island Red and Australorp.

Fowl Manure:

Four to six chickens will produce between 60-90 kg (120-200 lb) of manure per year. In addition, the scratching litter from the pen is excellent for top-dressing or incorporating into the garden soil.

Poultry manure also may be used to make liquid manure, or it can be added to the compost heap.

EXTRACTION OF OILS & ESSENCES 8

The 'life force' of all plants is contained in volatile oil. This valuable essence can be extracted in different ways: distillation and enfleurage (herbs and flowers) to give you a pure essential oil; by the use of a carrier oil; in an oil press (seed oils); in alcohol or vinegar; or by steeping or boiling the plant material.

This chapter deals with the different methods of extraction, including individual instructions for particular oils. All procedures are easily carried out within the home and are uncomplicated and simple.

HERB AND FLOWER OILS

Sun Distillation

Sweet scented flowers, herbs and fruits will yield their aromatic oils by this process. Place fresh flower petals or herbs in a large, wide-mouthed glass jar and cover with distilled water. (If using fruit, tear and bruise the skin first.) Seal the jar with plastic cling-film, ensuring that it is air-tight, and leave where it will receive hot sun every day.

When a thin film of oil appears, gently lift it off with cotton wool and squeeze it into a small glass bottle. Seal the bottle tightly, reseal the distilling jar and continue the process until no more oil appears.

Essential Oil of Marigold (Calendula) Tightly pack flower heads into a broad-necked jar that has a tight-fitting lid. Leave in the sun for several days until an oily orange fluid appears on the bottom of the jar. This is pure calendula oil.

Repeat the process until you have collected sufficient oil.

Kettle Distillation

This is a simple, yet effective method for producing small quantities of essential oil. It requires no elaborate equipment and is easy and inexpensive to make. You will need a large, old-fashioned type kettle that is boiled on the stove, a length of rubber hose, a short length of glass tubing, a large shallow basin, and a collection flask (glass jar or jug will suffice).

Put 500 g (9 oz) of fresh, sweet-scented flower petals or herbs in the kettle, half fill with distilled water, replace the lid and seal with plasticine or Blu-tack so that no steam can escape. Cut the rubber hose in two, so that one length is three times longer than the other, insert the glass tubing between the two lengths of tubing and attach the shortest end to the spout of the kettle. Put the kettle on the stove over a low heat. Place the basin so that it is higher than the kettle, fill it with ice-cold water and ice cubes, and run the hose through it, letting it hang down into the collection flask. Let the water in the kettle simmer until it has completely evaporated; the glass tube will allow you to observe when no more steam is passing through. Throughout the process keep the basin topped up with iced water or ice cubes.

In the collection flask two layers will be seen: the distilled water; with the pure essence floating on top of it. Gently lift off the fragrant oil with cotton wool and squeeze it into a small glass bottle.

Preparation

Before commencing the steam distillation process note that:

- dry, hard or fibrous substances should be bruised or macerated in water before distilling
- salted flowers and herbs are superior to fresh ones, as they reach the full development of their aromatic properties in a shorter time. To salt, spread the appropriate scented flowers or herbs on a shallow, flat tray and sprinkle a small quantity of salt over them. Place the salted material in a glass jar and cover with distilled water, seal, and allow to stand overnight. Add the contents of the jar to your kettle still, adjusting the distilled water if necessary
- distilled fragrant oil may have a smoky odour at first. Exposure for a short time to the air will remove this, after which it should be kept in tightly sealed bottles
- when distilling, ensure that the condensing tube is kept cold. The condensing steam must drip, not run

Yield proportion of some herbs:

Rose Petals	500 g (1 lb) will yield 0.02 g ($^1/_{100}$ dram)
Geranium (all types)	500 g (1 lb) will yield 0.5 g ($^1/_4$ dram)
Mints	500 g (1 lb) will yield 0.375 g ($^1/_6$ dram)
Lavender	500 g (1 lb) will yield 2.5 g ($1^1/_4$ dram)
Eucalyptus	500 g (1 lb) will yield 2.5 g ($1^1/_4$ dram)

Enfleurage

This process can be used to extract a pure essential oil or extract the essence using a carrier oil, such as olive, sunflower or other seed oil.

Essential Oil Choose fresh herbs or flower petals in the morning, after the dew has evaporated and they are at their most fragrant. Select only perfect specimens, discarding any damaged ones and keeping different types of herbs separate.

Place a layer of petals or herb leaves in the bottom of a small ceramic casserole or jar (never use glass or metal). Sprinkle a thin layer of coarse salt over them, and then repeat this procedure until the vessel is full.

Put the lid in place and seal tightly with plasticine or Blu-tack. Leave undisturbed for a month in a cool, dark cupboard.

After a month strain through muslin cloth into a glass jar, squeezing all liquid from the herbs. Seal the jar and leave where it will receive plenty of sunlight for 6 weeks so that it will settle. You will now have a concentrated fragrant oil. If you wish, strain one more time through filter paper.

Using a Carrier Oil Again choose fresh herbs and flower petals as before. Spread the selected petals and herb leaves on a shallow, flat tray and sprinkle a small quantity of non-iodised salt over them. Place a layer of the herbs in a wide-mouthed jar and then a layer of cotton wool, combed out thinly and soaked in a suitable carrier oil, such as almond, olive or sunflower oil. Repeat this procedure of alternating layers until the jar is full. Place the jar on top of a sheet of clear plastic and tie the plastic tightly over the top, enclosing the jar. Leave in a sunny spot for at least 15 days, then squeeze the fragrant oil from the whole mass. Strain and store in a tightly capped bottle.

Alternative Carrier Oil Method With this method you can extract the essence of either fresh or dried herbs.

Put two handfuls of fresh or dried flower petals or herb leaves (bruise fresh leaves) in a large glass jar and cover with ½ litre (1 pint) of carrier oil (any non-fragrant seed oil will suffice). Seal tightly, and leave in a sunny spot for 3-4 weeks, or longer if necessary, and shake the contents vigorously every second day.

Strain into a tightly corked bottle.

Maceration This method is especially suitable for roses, but will work well on all flowers and aromatic herbs. Place the flowers or herbs in a stainless steel or enamel pan, cover with an odourless vegetable oil and heat to 65°C (150°F). Remove from heat, cool, strain and store in a tightly sealed bottle.

Garlic Oil Chop up 6-12 garlic cloves (depending upon desired strength) and add to ½ litre (1 pint) of odourless vegetable oil in a glass jar. Seal tightly and allow to stand for 10-14 days. Strain into glass bottles and cork tightly.

Keeping Herb and Flower Oils Essential oils should always be stored in small, dark-coloured, air-tight glass bottles, away from heat and light. Never keep mixed oils, and especially those made with a carrier oil base, any longer than two months as they begin to oxidise immediately they are blended.

Seed and Vegetable Oils

There are different ways of extracting seed oils, the most common being to extract the oil with a seed press. Those plants best suited to this method are flax (linseed oil), jojoba, olive, sesame, sunflower and walnut (great to use in salads).

Before pressing, seeds are crushed or cracked and then made into cheeses (packs of crushed seeds wrapped in muslin cloth), piled on top of each other in the press, and then pressed to extract the oil.

If the cracked seeds are heated first in water until it boils, you will get more oil, although the quality is slightly less.

Preparation of seeds

flax	crack ripe pods with a hammer or pestle and mortar

jojoba	crack seeds as you would for flax
olive	gather ripe, nearly ripe and those laying on the ground. Olives left on the tree until they are fully ripe — during mid-winter and when black — will yield far more oil; about 10 per cent of their weight in oil.
	Crush in a coffee grinder or use a hammer. When fruit and stones are thoroughly crushed they are ready to be pressed
sesame	grind seeds thoroughly in a blender then press
sunflower	pulverise seeds with a hammer or pestle and mortar or process them in a blender
walnut	select only fully ripe nuts and leave to mature for at least a month after harvesting before pressing.
	Reduce to a pulp in a blender and pack into cheeses and press

If you are going to extract your own seed oils on a regular basis it would be worth investing in some form of crushing mill. There are a number of small crushers available designed primarily for crushing bone or rock that will suffice.

Making Your Own Oil Press

An oil press can be made with a suitable length of 150 mm (6″) circular steel pipe, a flat length of steel bar and an old wood-vice tightening bar. The following diagram details the simplicity of its construction, and should be within the capabilities of most people. If, however, you are not skilled in the use of an electric welder, or do not have access to one, your local garage or engineering shop should weld it together for a modest fee.

Using the Press

Once the seed packs are in the press, tighten the follower down, slowly applying pressure over 2-3 hours. Leave to drip overnight,

OIL PRESS

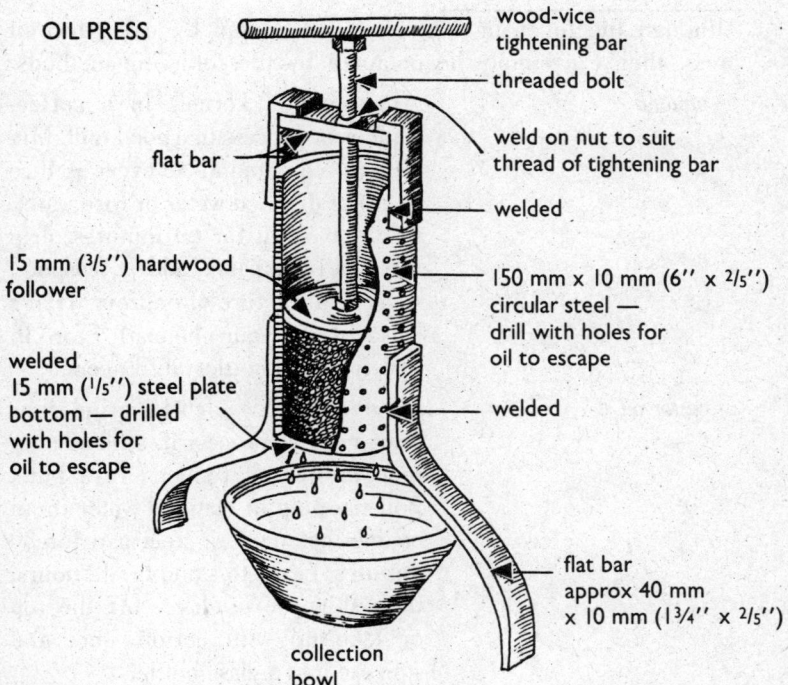

wood-vice
tightening bar

threaded bolt

weld on nut to suit
thread of tightening bar

welded

flat bar

15 mm (³/₅″) hardwood
follower

150 mm x 10 mm (6″ x ²/₅″)
circular steel —
drill with holes for
oil to escape

welded
15 mm (¹/₅″) steel plate
bottom — drilled
with holes for
oil to escape

welded

flat bar
approx 40 mm
x 10 mm (1³/₄″ x ²/₅″)

collection
bowl

then press again the next morning until no more oil is given up.

Remove the oil and place a clean collection container beneath the press. Release the follower, pour boiling water onto the cheeses — just enough to make them wet — and press again. Continue until the liquid coming out becomes clear: mostly water and no oil. Leave overnight to allow the water and oil to separate, and then siphon off the oil.

This second process gives you a slightly inferior oil, but can still be used in any of the recipes in this book, or blended with the first pressing.

When pressing olive oil, carry out both procedures, using the second method when no more blackish juice comes through. After siphoning, strain the oil through cotton wool that has been wrapped in a clean tea-towel; otherwise the oil will be slightly bitter tasting.

Store oils in air-tight bottles away from heat and light.

Extraction of Selected Oils
_____

Although the following oils can be extracted by using the oil press, they can simply be obtained by the following methods:

almond
grind almond kernels in a coffee-grinder or process in a good mill. Mix 20 g (¾ oz) almonds to every ½ litre (1 pt) of distilled water or rose water. Leave to steep for 60 minutes, drip through filter paper, add ½ teaspoon sugar and tincture of benzoin (friar's balsam from your chemist). Store in tightly sealed bottles until needed.

castor oil
grind seeds in a coffee-grinder or flour mill, or process in a blender or food mill, until reduced to a pulp. Boil the pulp in distilled water in an enamel or stainless steel pan for 30 minutes. Leave to stand for 12 hours, then skin the oil layer off the top or take up with cotton wool and squeeze into a glass bottle.

peanut oil
grind peanuts in a blender or coffee-grinder and add the pulp to a bowl of boiling water. Allow to steep overnight, then next morning skim off the oil floating on top of the water.

safflower
harvest safflower seeds when slightly under-ripe to ensure maximum oil yield. Reduce the seeds to a pulp in a blender and boil in water for 30 minutes. Leave to steep overnight then gently take up oil by skimming or with cotton wool.

Flower and Herb Waters

The simplest method of extracting the benefits of herb oils is that of steeping fresh or dried herbs in boiling water, and is known as a 'herbal infusion'. However, the bark and roots of a plant will not yield their properties by this process and must be extracted by a method called a 'decoction'. The delicate scent of flowers needs to be treated differently again, and can be obtained by a variety of processes.

Herbal Infusion

Place the selected herbs in a large ceramic bowl and pour boiling water over them. Cover and steep for 12 hours (or overnight), unless otherwise directed, then strain through muslin cloth and add required amount to the recipe.

The following proportions should apply:
> fresh herbs 3-4 tablespoons (4-5 US tbsp) or dried herbs
> 1-2 teaspoons to 300 ml (10 fl oz) of boiling water

Herbal infusions can be made from one herb or a mixture of herbs, depending upon the requirements of a recipe or your particular needs.

Usually, most infusions will only last one or two days unless kept in a glass bottle in the refrigerator, when they last up to a week. To extend the keeping qualities of the infusions add 5-10 millilitres (1-2 tsp) of vodka to every 300-500 millilitres (10-16 fl oz) of infusion, or 3-4 drops of tincture of benzoin (friar's balsam) to every 300 millilitres (10 fl oz).

Decoction

Put the selected herbs in a stainless steel or enamel saucepan and add distilled water. Bring to the boil and then gently boil for 30 minutes (unless otherwise directed). Remove from heat, cover the pan, and allow to steep for 10 minutes. Strain through muslin cloth and add required amount to recipe.

The following proportions should apply:
> fresh herbs 3-4 tablespoons (4-5 US tbsp)
> or dried herbs
> 1-2 teaspoons to 300 ml (10 fl oz) of distilled water

Flower Water by Distillation

This method makes a very potent floral water and is especially suited for making rose water, lavender, marjoram, pennyroyal and basil waters. Using the kettle still, follow the same procedure for making essential oils, and instead of taking up the fragrant essence blend it with the distilled water until thoroughly emulsified.

Flower Water Decoction

This recipe will give you beautifully scented water from any fragrant flower.

Put 4 tablespoons (5⅓ US tbsp) of fresh flower petals in a saucepan and cover with 1½ cups of water. Bring to simmering point, cover and simmer for 30 minutes. Remove from heat, cool, strain through muslin cloth, and squeeze any remaining liquid from the flower petals.

If necessary, and depending on the type of flowers used, repeat the process for greater potency. Add fresh flowers to the liquid and top up if required.

Keeping Herb and Flower Waters

All herb and flower waters prepared using the preceding methods can have their keeping qualities extended. Lavender, due to its natural antiseptic properties, has good keeping qualities, while other herbs will keep quite well if vodka or tincture of benzoin is added as directed under Herbal Infusion.

Should you wish to make large quantities of fragrant water, and ensure that it will last indefinitely, add one part alcohol (surgical spirit if you can obtain it from a friendly chemist, if not, use vodka) to two parts fragrant water, stand for twelve hours, then drip through filter paper.

Distilled flower waters do not require preservation, since the blended oil and water will keep well without these additions for a considerable time.

Essential Oil and Water

Two drops of pure essential oil, extracted without a carrier oil medium, added to one litre (2 pints) of water will make a fragrant herb or flower water. Add it to the final clothes wash cycle, use it as an ironing spray, or include in your bath water.

Herbal Vinegar

Herbal vinegars have been used for centuries as bath additions and beauty tonics. They can also be used as the basis of a natural deodorant and facial toning lotion, and have a far more refreshing scent than floral waters made with alcohol, as well as softening both facial and body skin.

Their use is not only limited to skin care and toiletry products, but just as effectively can be used in many domestic household products and as a replacement for oil in summer salads.

Basic Recipe for Herbal Vinegar

> 1½ cups fresh flower petals or herbs

> 2 cups cider vinegar or white wine vinegar

Put the flower petals or herbs into a large, wide-mouthed glass bottle. Gently warm the vinegar, pour into the bottle, seal tightly, and leave where it will receive plenty of hot sun for 2 weeks. Shake the contents every day. Strain the vinegar and store in tightly capped bottles.

If the scent is not strong enough, repeat the process with a fresh batch of flowers or herbs.

Uses of Vinegar

Apart from the various uses already discussed throughout this book, herbal vinegars can also be used to soothe a throbbing temple after exposure to the sun, refresh a sick room, and act as an astringent face wash.

Sun Exposure　After exposure to the hot sun, dab lavender, rose or lemon verbena vinegar behind the ears and on the temples and forehead.

Face Wash

> 2 tbsp (2⅔ US tbsp) lavender vinegar

> 1 cup rose water

Mix thoroughly and store in tightly capped bottles. Apply directly to face.

Other herbs and flowers suitable for a face wash:

basil	*jasmine*
dill	*lemon verbena*
sweet scented geranium leaves	*violet*
honeysuckle	

Refresh a Sick Room　Soak a small sponge in lavender vinegar and leave it in a dish beside the bed.

Herbs Suitable for Salad Vinegars
***

Try using herbs such as tarragon, garlic, dill, rosemary, marjoram, oregano, basil, orange-mint, thyme, caraway, or any other favourite herb.

Prepare the vinegar as you would in the basic recipe, except garlic and herb seeds. Strain the contents into old salad dressing bottles or similar fancy bottles for the vinegars you may want to give as gifts; old cruets can be used for an elegant touch. Add a spray of the herb before sealing, making sure there are no frayed leaves or hidden bugs, as any imperfections will be magnified by the vinegar in the bottle.

Vinegar from Herb Seeds

Bruise the seeds with a pestle and mortar and allow two tablespoons (2⅔ US tbsp) of seeds to every 1 litre (2 pt) of vinegar. Then prepare as for basic recipe.

Garlic Vinegar

Put garlic cloves into vinegar and leave to steep for 24 hours. Strain into bottles and seal tightly.

Drying Herbs

Since many of the annual herbs are not available all year, they should be gathered fresh when in season and preserved for future use. A simple way to preserve herbs is to dry them, ensuring that they can be used all year round. So that they retain their colour and qualities, herbs have to be picked at the right moment and dried immediately afterwards.

- gather herbs in the early morning when they are dry but before the sun has had a chance to draw out and disperse their volatile oils
- do not pick herbs while they are still damp with dew or after rain
- use a sharp knife unless harvesting chives, which are best cut with a pair of sharp scissors
- spread picked herbs out thinly on trays
- do not collect more than you can immediately dry, as they will only deteriorate and lose their essential oil
- avoid cross-flavouring by keeping different herbs separate

■ handle as little as possible as herbs will bruise easily

When gathering herbs it is also important to know what and when to harvest.

flowers	select only those which are unblemished and pick when they are fully open
leaves	should be left attached to the stem and imperfect ones discarded. Volatile oils are at their peak just before the herb plant flowers, and leaves should be picked then
seeds	cut whole heads once they have turned brown and seeds are ready to fall. Leave on the stem
roots	horseradish may be harvested anytime; all other herbs are gathered in autumn

Because herbs contain about 80 per cent water, the object of drying after harvesting is to remove this water without delay to prevent the loss of their valuable and beneficial properties. They should be dried quickly with an even, low warmth — no higher than 32°C (90°F) nor lower than 21°C (70°F) — and away from direct sunlight or wind.

Air Drying

This requires no special arrangements: simply tie the herbs and flowers in bunches and hang them upside down in a dry, airy place, or spread them out thinly on net-covered drying trays and keep in a warm, well-ventilated place. Usually, herbs will take 4-12 days to dry, sometimes longer.

leaves	are dry when brittle, but will not shatter
flowers	are ready when petals feel dry and slightly crisp
roots	should be dried right through with a soft centre

| seeds | after being removed from seedheads should be placed in the sun for a few hours |

Solar Drying

Herbs can be dried successfully in a solar drying cabinet, the plans of which are in chapter 1.

Spread herbs or flowers thinly on trays, made from a wooden frame and covered with muslin cloth, and stack in the cabinet. The drying space should only smell slightly of herbs. From time to time turn the herbs and do not add any fresh ones until each batch is dry. Doing so will only add more humidity to the air.

Generally, herbs are ready once really brittle in leaf and stalk but still green in colour. To ensure even drying it may be advisable to change the top and bottom trays during the process. However, be sure to do this without interruption.

Remove herbs each evening until dry to avoid moisture re-occurring.

Oven Drying

A less satisfactory method, since temperatures must be maintained below 32°C (90°F). Spread the herbs on trays and leave the oven door ajar to ensure ventilation.

Microwave Oven Drying

Most herbs, especially parsley, which does not respond well to air drying, can be dried successfully in a microwave oven. It is advisable to do a few trial runs first to determine the best drying time. Turn the oven on to full power and lay the washed herb on two layers of absorbent paper on top of an ovenproof tray. Usually it should not take longer than four minutes to dry each batch. Herbs with delicate, feathery leaves such as fennel and dill weed, do not respond to this type of drying technique.

Storing Dried Herbs

When dry, herbs are rubbed gently by hand to discard the hard stalks — not too finely or they will lose their fragrance — and stored in clean air-tight glass containers. If storing in metal bins, first place them in a linen or cotton bag. All containers should be kept in a cool, dark place so that the herbs retain their fragrance.

The following should be observed when storing herbs:

- thyme, sage and rosemary can be left on the stalk
- seeds and flower-heads should be placed immediately into air-tight containers
- roots should be ground like coffee beans in a grinder or blender, and only stored in air-tight glass containers

If at any time the container shows signs of moisture on its inside, the herbs weren't correctly dried. Remove the herbs and place them on a sheet of plain paper and allow further drying time.

At the most, dried herbs will only last a year, so always remember to replace them. An accurate check is to date your containers.

INDEX

FOR YOUR NOTES

FOR YOUR NOTES

FOR YOUR NOTES

Other books by the author:

Healing from the Garden
A beautifully illustrated medicinal herbal that describes the
healing properties of all the herbs and plants listed in the
book. It details which parts of the herb should be used for
which ailments and gives clear instructions about how
they should be prepared.

Beauty from the Garden
A comprehensive guide to making completely natural soaps,
lotions and creams to pamper the skin, and gentle shampoos
and conditioners to care for your hair. It explains how to
extract essential oils, how to dry herbs and fragrant plants,
and how to use oils for aromatherapy.